Operating an Amateur Radio Station

AF712483

Contributors:

Paul L. Rinaldo, W4RI
John Nelson, W1GNC
Robert Halprin, K1XA
Stan Horzepa, WA1LOU
Kirk Kleinschmidt, NTØZ
David Newkirk, AK7M
Richard K. Palm, K1CE
Mark Wilson, AA2Z
Larry Wolfgang, WA3VIL
Jim Cain, K1TN
Steve Ewald, WA4CMS

Published by
The American Radio Relay League
225 Main Street
Newington, Connecticut 06111

FSD-900

Printed in USA

64th Edition

ISBN: 0-87259-226-X

FOREWORD

Operating a station is the essence of Amateur Radio. An Amateur Radio license is your passport to virtually every country in the world—without leaving your station. Your ticket also permits you to communicate with other nearby hams who can help you grow in knowledge and provide lifelong friendships.

At its simplest, operating can be almost as easy as talking on the telephone. But there's much to learn if you want to take advantage of the full range of operating opportunities. These include delights such as mastering Morse code to the degree that it becomes a second language, connecting your computer to a growing worldwide packet-radio network, amateur television, and even communicating through satellites.

The four cornerstones of successful operating are: knowledge, skill, dedication and courtesy. This booklet is intended to give you the basic knowledge needed for operating an Amateur Radio station. Skill is a product of operating ability and repeated practice so that the mechanics become almost second nature. Dedication involves spending the time to learn, then applying that to your enjoyment of this rich hobby and serving the public. Courtesy in Amateur Radio goes beyond just being considerate and extends to being helpful to others.

This booklet is an introduction to operating. *The ARRL Operating Manual* is the complete resource book on operating an Amateur Radio station, and is sold at radio stores and directly from ARRL Headquarters. The *Public Service Communications Manual* explains traffic-handling and emergency net operations, and is available free to League members on request (please include a 6- × 9-inch self-addressed, stamped envelope with two units of first-class postage).

See you on the air. 73,

David Sumner, K1ZZ
Executive Vice President

The Amateur's Code

The Radio Amateur is:

CONSIDERATE . . . never knowingly operates in such a way as to lessen the pleasure of others.

LOYAL . . . offers loyalty, encouragement and support to other amateurs, local clubs, and the American Radio Relay League, through which Amateur Radio in the United States is represented nationally and internationally.

PROGRESSIVE . . . with knowledge abreast of science, a well-built and efficient station and operation above reproach.

FRIENDLY . . . slow and patient operating when requested; friendly advice and counsel to the beginner; kindly assistance, cooperation and consideration for the interests of others. These are the hallmarks of the amateur spirit.

BALANCED . . . radio is an avocation, never interfering with duties owed to family, job, school, or community.

PATRIOTIC . . . station and skill always ready for service to country and community.

—The original Amateur's Code was written by Paul M. Segal, W9EEA, in 1928.—

TABLE OF CONTENTS

Introduction

SAFETY FIRST

Ham radio is an extremely safe hobby provided that you use common sense as you would working around any electrical or mechanical equipment. You need to follow simple safety procedures since the misuse of radio or other electrical equipment can cause death or permanent injury. Before you install, repair or use any equipment, become familiar with the detailed safety rules published in *Tune in the World with Ham Radio* or *The ARRL Handbook*.

Know your equipment. Never work alone on the "innards" of radio equipment, or when erecting antennas or using power tools.

Take precautions against prolonged body exposure to radiated radio-frequency energy, particularly in the VHF and UHF bands. Switch to safety!

CHOOSING YOUR FIRST RIG

Choosing your first ham station might seem a bit confusing. Should you buy a new radio—with lots of pushbuttons, digital frequency readout and fancy features—or an older, "barebones" rig that still offers good performance? Will a transceiver suit your needs, or can you manage a separate receiver and transmitter? Tubes or solid state? Tune or no-tune? Big, small or in between? Although the questions seem endless, choosing your first rig doesn't have to be a chore.

In times past, most Novice hams had a radio mentor—an Elmer—who first fired their interest in ham radio and was there at every turn to answer questions and provide assistance. With the help of such an Elmer, choosing a rig is a simple task! Most Elmers have operated many different rigs over the years, and they won't hesitate to recommend several alternatives appropriate for any budget.

If you are fortunate enough to have such an Elmer, you probably have several possible rigs picked out. If you don't have a personal ham-radio

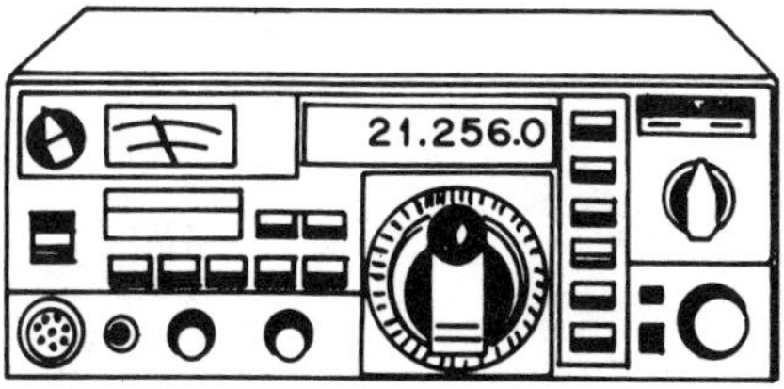

helper, don't despair. Most hams love to talk about equipment, so it shouldn't be too difficult to find a few local hams who'll be more than happy to help you in your search for a first rig. If you're lucky, you'll be invited to stop by their shacks and spin the dials on their equipment for yourself. That's another thing hams like to do—show off their shacks!

Your local radio club is a good place to meet other hams. If you can't find any in your area, send an SASE to the Field Services Department at ARRL HQ, and we'll send you information on the clubs that meet in your area.

If you are fortunate enough to live near a ham radio showroom, you can get a first-hand look at many of the latest rigs. Many dealers have a supply of used gear on hand as well. Don't let the high prices of today's new radios scare you away. If your budget won't allow you to purchase a brand new transceiver, there are plenty of used rigs that will get the job done—at a price you can afford.

A good way to become familiar with the features (and drawbacks) of used and new rigs alike is the Product Review Column in *QST*. Many, many rigs have been reviewed there over the years. The reviews detail performance specifications, features and controls, as well as provide comments about the rig's on-air performance. Look in the *QST* annual index, located at the end of each December issue, to see which rigs were reviewed that year. A cumulative Product Review index going back to 1976 is available from the ARRL Technical Department Secretary (send $3 and an SASE). Radio clubs and your

local library often have back issues of *QST*.

Two ARRL publications address the question of choosing a first rig in greater detail. One is *Tune in the World With Ham Radio*; the other is *The ARRL Operating Manual*. Both books explore these subjects in detail: solid-state and older tube-type rigs; transceivers and separate transmitters and receivers; digital and analog frequency displays; and features and accessories such as audio filters, speech processors, among others.

When you've sorted out the alternatives and finally decided which rig (or rigs) to buy, the question of *where* to buy still remains. Some possibilities include local hams, local dealers, flea markets, mail order, Ham Ads in *QST* and classified ads in other publications. Your local dealer will probably offer the best service, although the prices may be higher than at comparable mail-order outlets. Gear obtained through flea markets, local hams, and classified ads usually costs less than similar equipment purchased from an established dealer, but there's usually no guarantee as to the condition of the equipment. *Caveat Emptor* (buyer beware) still applies—even in ham radio. And, you generally get what you pay for.

As you gain experience in Amateur Radio, the process of choosing a radio will become less mysterious—but no less intense. New models are introduced every year and their features are hotly debated, on the air and off. Because of this, most hams trade their gear in several times throughout the course of their ham radio career. You probably will, too.

If you stay with Amateur Radio long enough, someday you'll be somebody's Elmer—helping a ham radio newcomer find that first rig. . . "I've got an old rig I use as a backup—it's certainly not new, but I'm sure it'll work fine if you'd like to try it. . ."

ANTENNAS FOR YOUR STATION

The subject of antennas holds a great deal of mystery for an inexperienced ham (and many experienced ones as well). There seems to be an endless string of questions about these magic devices. How long should an antenna be? How high does it have to be? What kind of feed line should I use? What do I use to hold the antenna up in the air? What if the SWR is too high? What size wire should I use to build the antenna? Should the wire be insulated or not? What if I can't put up an outside antenna? What can I do if my yard isn't big enough to stretch a dipole out straight?

Before you become too worried about all these antenna questions, you need to understand the basic operating principle of all antennas. ***Any*** *antenna at all will work much better than* ***no*** *antenna!*

So now you know: Put up anything at at all, no matter how significantly different it is from the "ideal, theoretical antenna." You will have far more fun using it than you will have sitting there worrying about what kind of antenna to try! While you are enjoying ham radio by contacting other hams on the air, you may wonder what antenna to try next or how to improve the system you have now, but don't fret so much about what to put up that you never try any antenna. During your on-the-air operating, you can ask other hams for new antenna ideas.

What is the simplest antenna for an inexperienced ham to put up? You might be tempted to buy a commercial antenna, thinking you will be incapable of building your own. Resist that temptation and try building a dipole antenna. Build your antenna from materials you have available. Just about any type of wire will do, if it is strong enough to keep from breaking with the weight of the wire, feed line and insulators.

For supports, you can use whatever is

convenient. A couple of tall trees should work well. You can run one end to a corner of your house or garage. You can support the center of the antenna in a tall tree, with both ends sloping toward the ground. This forms an inverted-V antenna. You can also tie one end in a tall tree and slope the entire antenna toward the ground. If you don't have enough room to put the dipole up straight, try bending the ends to form a Z. If you can't get the antenna outside, you can even put it up in your attic (or in another room in your house).

What should you use for feed line? Many hams use 50 or 72-ohm coaxial cable, but parallel-conductor line will have less loss. You can buy or build open-wire feed line or you can even use TV-type twin lead. You will need some type of impedance-matching network, or Transmatch if you use parallel-conductor feed line.

This section of *Operating an Amateur Radio Station* isn't intended to provide detailed construction information. The ARRL publishes a number of books that contain plenty of details on basic antennas. You should have one or more of those books as a reference. *Tune In The World With Ham Radio* contains complete plans for a number of simple antennas. *The Novice Antenna Notebook* and *W1FB's Antenna Notebook*, both by Doug DeMaw, contain lots of practical antenna information. *The ARRL Antenna Compendium* is a compilation of previously unpublished antenna articles. Finally, there's *The ARRL Antenna Book*, which contains just about all the information you'll ever need to know about antenna theory and construction.

You might also want to try a quarter-wavelength vertical antenna. You can build such an antenna using lightweight aluminum tubing, copper tubing or wire. You'll need something to support a wire vertical, and that can be a wooden mast, a fiberglass stick (such as from a bicycle safety flag) or other materials. You can string a wire along a tall, straight tree trunk, and this can be a very effective support for an 80 or 40-meter vertical antenna. You'll have to con-

struct a system of ground radials under your vertical antenna. How many radials should you use? Remember the fundamental principle of antennas? Start with four radials and experiment by adding more later.

Here is a table of lengths for ½-wavelength dipole and ¼-wavelength vertical antennas for the 160 to 10-meter amateur bands. Use these lengths as a rough guide to the length for your antenna. It's probably a good idea to start with the antenna a bit longer than the length given here. The dipole length is cut in the middle, and the feed-line connection is made there.

Lengths for ½-λ Dipole and ¼-λ Vertical Antennas for the Amateur HF Bands

Band (meters)	*Design Frequency (MHz)*	*Dipole (feet)*	*Vertical (feet)*
160	1.9	246	123
80	3.7	126	63
40	7.2	65	32.5
30	10.12	46	23
20	14.225	33	16.5
17	18.1	27	13.5
15	21.15	22	11
12	24.94	19	9.5
10	28.3	16.5	8.5

When you are ready to put your antenna into the air, there are several things you should be careful about. First, do not put up any antenna where it can fall and come into contact with an electrical power line. Do not attempt to string a dipole over the top of a power wire under any circumstances. Another thing you should be careful about is in regard to climbing trees, towers or ladders to erect any type of antenna. Be sure to take extra safety precautions when working above ground level. Don't let a careless slip ruin your ham radio fun!

Remember our basic antenna fundamental: *Any antenna at all will work much better than no antenna!* So put up *some* type of antenna, and enjoy operating ham radio. Good luck.

TECHNICAL INFORMATION

Got a technical question concerning your rig or antenna? Or, troubled by a nasty RFI problem? Well, you're not alone. Hundreds of old-timers and newcomers alike have questions and problems just like yours.

Good news! There's help waiting for you right around the corner. The ARRL Technical Information Service is administered through an extensive network of volunteers around the country, in your state, section, county and town. They are League Technical Coordinators and special Assistants who can help you resolve your problems and answer your questions. They are experts on most technical topics and have a keen appreciation and understanding of the difficulty of radio frequency interference problems.

Contact them! You can find the name, address and phone number of your local Technical Coordinator by contacting your Section Manager (his or her name, address and phone are listed on page 8 of every *QST*).

When writing to your TC, keep the following guidelines in mind:

1) Type or write the question clearly and to the point; adequately describe the problem, action already taken, and question.

2) Include your name, call and license class, and provide a business size self-addressed, stamped envelope (SASE).

W1AW

The Headquarters station of the American Radio Relay League, W1AW, transmits daily on-the-air code practice and bulletins of interest to hams. It's a good habit to copy W1AW CW transmissions to improve your code speed and keep up with the latest happenings in Amateur Radio.

The W1AW schedule can be found in *Tune in the World with Ham Radio* and in *QST*.

TIME

Most references to time of day in Amateur Radio are in terms of Coordinated Universal Time (UTC). It's a good idea to have a station clock set to UTC so you don't have to waste time converting from local time to UTC for log entries.

Converting Local Time to UTC

UTC	*AST/ EDT*	*EST/ CDT*	*CST/ MDT*	*MST/ PDT*	*PST*
0000	2000	1900	1800	1700	1600
0100	2100	2000	1900	1800	1700
0200	2200	2100	2000	1900	1800
0300	2300	2200	2100	2000	1900
0400	0000	2300	2200	2100	2000
0500	0100	0000	2300	2200	2100
0600	0200	0100	0000	2300	2200
0700	0300	0200	0100	0000	2300
0800	0400	0300	0200	0100	0000
0900	0500	0400	0300	0200	0100
1000	0600	0500	0400	0300	0200
1100	0700	0600	0500	0400	0300
1200	0800	0700	0600	0500	0400
1300	0900	0800	0700	0600	0500
1400	1000	0900	0800	0700	0600
1500	1100	1000	0900	0800	0700
1600	1200	1100	1000	0900	0800
1700	1300	1200	1100	1000	0900
1800	1400	1300	1200	1100	1000
1900	1500	1400	1300	1200	1100
2000	1600	1500	1400	1300	1200
2100	1700	1600	1500	1400	1300
2200	1800	1700	1600	1500	1400
2300	1900	1800	1700	1600	1500

Basic Operating

Standard radio operating procedures make communicating easier and more efficient. Such procedures include the use of Q signals (QRM, QSL and so on), procedural signals (*prosigns* such as K and $\overline{\text{AR}}$) and abbreviations that are used to convey commonly accepted meanings yet save time. Many commonly used Q signals, prosigns and abbreviations appear in the appendix. A line over two letters means the letters are sent without a space between them. $\overline{\text{AR}}$ is sent as *didadidadit* and not *didah didadit*. Just as in ordinary conversations, there is a general order to radio contacts. For instance you don't say "goodbye" before you say "hello." Nonstandard use of these signals, or sending things out of order, invariably causes confusion. Amateurs are not bound by most of these conventions in any legal way. It just makes good sense to send what will be easily understood by other amateurs.

Like the English language, however, Amateur Radio procedures evolve over time, usually changing by common usage and sometimes by agreement. You may hear wide deviation from the procedures described in this booklet in your on-the-air activities. We'll give you the "normal" way things are sent and let you pick up the on-the-air slang on your own!

LISTENING

Skillful radio operators listen about 90% of the time and transmit only 10%. Before you first get on the air, listen 100% of the time, or until you feel you know how the better operators conduct themselves. Copy the best and forget the rest!

A cardinal rule in Amateur Radio is to listen before transmitting on any frequency. At best, transmitting without listening is a rude interruption of someone else's conversation. At worst, it could interfere with emergency communications in progress.

Listen first, then ask "Is this frequency in use?" if operating phone. If it is in use, the response is a simple "Yes." On CW, send QRL?; the affirmative response is C.

Packet radio protocols have a listen-before-transmit feature built in. Radio systems that operate automati-

cally have circuits that listen before transmitting, as well as watchdog timers to turn the transmitter off the air after a maximum of three minutes in the case of malfunction.

Once you are finished with a frequency, let others know. On phone use the word "Clear"; on CW, use the prosign $\overline{SK}$.

IDENTIFICATION

FCC rules require that each station identify itself at least every 10 minutes. There is more to the ID rules, so it's a good idea to read the FCC rules (published in *The FCC Rule Book)* before operating the first time.

Most hams practice sending their newly assigned call sign to themselves until they master it on CW and get used to the phonetics for each letter on voice. The International Telecommunication Union phonetic alphabet is as follows:

A—Alfa	(**AL** FAH)
B—Bravo	(**BRAH** VOH)
C—Charlie	(**CHAR** LEE)
D—Delta	(**DELL** TAH)
E—Echo	(**ECK** OH)
F—Foxtrot	(**FOKS** TROT)
G—GOLF	(GOLF)
H—Hotel	(HOH **TELL**)
I—India	(**IN** DEE AH)
J—Juliett	(**JEW** LEE **ETT**)
K—Kilo	(**KEY** LOH)
L—Lima	(**LEE** MAH)
M—Mike	(Mike)
N—November	(NO **VEM** BER)
O—Oscar	(**OSS** CAH)
P—Papa	(**PA** PA)
Q—Quebec	(KWEE **BECK**)
R—Romeo	(**ROH** ME OH)
S—Sierra	(SEE **AIR** RAH)
T—Tango	(**TANG** GO)
U—Uniform	(**YOU** NEE FORM)
V—Victor	(**VIK** TAH)
W—Whiskey	(**WISS** KEY)
X—X-Ray	(**ECKS** RAY)
Y—Yankee	(**YANG** KEY)
Z—Zulu	(**ZOO** LOO)

Note: The **boldface** syllables are emphasized.

On most HF CW and phone contacts, it is customary to send the call signs of both stations in communication at the beginning and end of each transmission. The exceptions are called break-in operation, where call signs are not sent during each transmission. Break-in operation is the norm for contests and ragchewing contacts between experienced operators.

When using an FM repeater, be sure to identify your station during your first transmission. A sure way to have your ear chewed on is to "kerchunk" (key up briefly without identifying) a repeater that has been recently bothered by a frequent kerchunker.

ESTABLISHING CONTACT

At first, rather than calling CQ yourself, it is best to *listen* patiently until you hear someone calling CQ, and call *them*. With most modern transceivers, once you have the other station tuned in correctly, the transmitter frequency will be automatically adjusted to the same frequency as the other station. If you have separate receiver and transmitter, tune ("spot") your VFO frequency until you hear the spotting signal your receiver. Don't spot your transmitting frequency by putting a signal on the air.

Calling CQ

Assume that W1AW is calling CQ, meaning "I wish to contact *any* amateur station." W1AW would do

so by listening for a clear frequency. If the frequency appears to be clear, W1AW asks, "Is the frequency in use?" On CW, W1AW sends QRL?. If no one is using the frequency, W1AW then sends:

CW: CQ CQ CQ DE W1AW W1AW W1AW K

Phone: CQ CQ CQ THIS IS WHISKEY ONE ALFA WHISKEY, OVER

If no one answers, this sequence may be repeated several times. Brief, repeated calls with frequent standbys are preferred. Long drawn-out calls simply clutter up the airwaves and discourage other stations from answering you.

On FM repeaters, most operators don't call CQ, but say something like "This is W1AW monitoring." It's the functional equivalent of calling CQ. There's nothing wrong with calling CQ on repeaters, but it's not the custom, and someone may go out of the way to let you know that. Keep calls on repeaters short; there may be others who would like to use the repeater (or "machine," as many hams call them).

Answering CQ

If your call sign is WB9ZZZ, you would answer W1AW's CQ as follows:

CW: W1AW W1AW DE WB9ZZZ WB9ZZZ $\overline{AR}$

Phone: W1AW W1AW THIS IS WHISKEY BRAVO NINE ZULU ZULU ZULU, OVER

In answer to your call, W1AW would send:

CW: WB9ZZZ DE W1AW R. . .

Phone: WB9ZZZ THIS IS W1AW ROGER. . .

No, W1AW's name isn't Roger; "roger" means that W1AW acknowledges receipt of WB9ZZZ's transmission. After the "R" or "Roger" conversation begins. But,

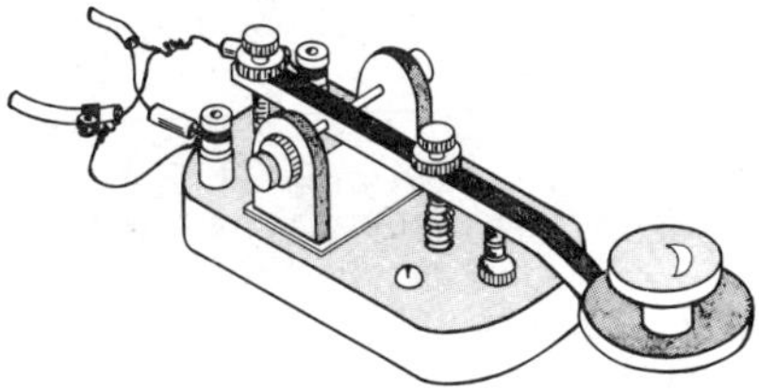

what do you do if someone answers you but you weren't able to copy the other station's call sign? It depends.

If you need a simple repeat of the other station's call sign, then ask "who is calling me?" as follows:

CW: QRZ? DE WB9ZZZ $\overline{AR}$

Phone: QRZ, THIS IS WHISKEY BRAVO NINE ZULU ZULU ZULU OVER

On phone, the general rule is to use ordinary English words wherever possible rather than Q signals. There are a few exceptions. For example, you could say "unknown station," but QRZ (using the British pronunciation "zed" which rhymes with red) is commonly used.

If you know that a W1 station called you but missed the rest of the call sign, transmit:

CW: QRZ W1? DE WB9ZZZ $\overline{AR}$

Phone: QRZ THE W1 STATION, THIS IS WHISKEY BRAVO NINE ZULU ZULU ZULU OVER

Sending the part of the call sign you received correctly is particularly useful whenever you hear more than one station calling you and you want to single out one station to answer you.

EXCHANGING RECEPTION REPORTS

After both stations have established contact by acknowledging with R or "Roger," the usual next step is to exchange signal reports. Generally, this is done using the *RST* system. On FM repeaters, it is customary to describe a strong signal with no noise as "full quieting," or to mention the amount of noise or fading.

RST stands for *readability* (a scale of 1 to 5), *signal strength* (1 to 9), and *tone* (1 to 9). On CW, a report of RST 367 would mean: "Your signals are readable with considerable difficulty, good strength, and a trace of modulation." On phone, the tone report is omitted.

The RST System

Readability

1—Unreadable.
2—Barely readable, occasional words distinguishable.
3—Readable with considerable difficulty.
4—Readable with practically no difficulty.
5—Perfectly readable.

Signal Strength

1—Faint signals barely perceptible.
2—Very weak signals.
3—Weak signals.
4—Fair signals.
5—Fairly good signals.
6—Good signals.
7—Moderately strong signals.
8—Strong signals.
9—Extremely strong signals.

Tone (Used on CW only)

1—Sixty-hertz ac or less, very rough and broad.
2—Very rough ac, very harsh and broad.
3—Rough ac tone, rectified but not filtered.
4—Rough note, some trace of filtering.
5—Filtered rectified ac but strongly ripple-modulated.
6—Filtered tone, definite trace of ripple modulation.
7—Near perfect tone, trace of ripple modulation.
8—Near pure tone, slight trace of modulation.
9—Perfect tone, no trace of ripple or modulation of any kind.

The tone report refers only to the purity of the signal and has no connection with its stability or freedom from clicks or chirps. If the signal has the characteristic steadiness of crystal control, an X may be added to the report (for example, RST 469X). If it has a chirp (frequency instability during each dot or dash), add a C; if key clicks are present, add a K.

For the most part, the addition of C, K or X has fallen into disuse because of the predominance of newer commercial radios without chirps or clicks when the equipment is working properly. C and K are still used to advise the other operator of an unusual problem.

The RST system has been the subject of misuse and controversy over the years. One problem is that many operators simply give the other station good news and may habitually send RST 599 or 579 to everyone regardless of the signal received. The options for the future of the RST system seem to be: (1) Use it, (2) abuse it, (3) modify it or (4) replace it. Time, discussion and evolving operating practice will tell.

WHERE, WHO?

After giving the RS(T) report, the next thing is to tell the other operator your location (QTH) and your name. The signal report, location and name can be transmitted quickly by sending something like:

CW: RST 579 IN NEWINGTON CT NAME PAUL

Phone: YOU'RE 57 IN NEWINGTON, CONNECTICUT. NAME IS PAUL.

Note the lack of punctuation on CW. Most CW operators use the double-dash ($\overline{BT}$) to separate phrases and sentences during longer transmissions or "exchanges."

Give the other station a chance to send similar information and copy it in your station log for future reference. If you are a bit absent minded, the log can be handy reference during the contact.

THE CONVERSATION

After the exchanges mentioned in the previous section, QSOs are a free art form. It's the same challenge as face-to-face communication: What do I say after hello? Maybe just good-bye? As strange as that might sound, that sort of thing happens by necessity in Amateur Radio QSOs during contests or when working a particularly rare DX station in a pileup.

Enter the gentle art of ragchewing. Some operators break the ice by describing their radio equipment, antenna and weather (WX). That's fairly usual stuff and not apt to engage the other operator in meaningful conversation. You'll be surprised at how well some foreign stations speak English even though it is not their native tongue. The key to good ragchewing is to discover shared interests. Chess? Family? Job? Horses? Travel?

Be sensitive to restrictions regarding freedom of speech in some countries. For this reason, and simply because it's good manners when talking to someone you don't know very well, it's best to avoid politics and religion when working foreign stations.

There are prohibitions against exchange of third-party traffic with other countries. What's a third party? The definition varies by country from "anyone other than the operator you're talking to" to "someone other than a ham." Just to be on the safe side, before asking someone to say "Hi" to your Aunt Minnie (that's third-party traffic) check whether that's legal. Check *The ARRL Operating Manual* or the latest *QST* Mini Directory. *QST* prints up-to-date listings of countries with which third-party traffic is permitted.

ENDING THE CONTACT

All good things must come to an end, including ham radio contacts. You may want to contact some different stations or perhaps you are going to end this particular operating session. On CW, the operator who wants to end the contact sends the Q signal "QRU" (I have nothing more for you) just before identifying. On phone this can be the statement "thanks for the QSO." If you want to go on and make more contacts you send:

CW: W1AW DE WB9ZZZ $\overline{\text{SK}}$

Phone: W1AW THIS IS WHISKEY BRAVO NINE ZULU ZULU ZULU CLEAR

$\overline{\text{SK}}$ on CW means end of contact. Sometimes on phone the final phrase is, "Clear and QRZ," or "Clear and listening for other calls." But if you are closing your station after this operating session (CL on CW means "closing down station"):

CW: W1AW DE WB9ZZZ CL

Phone: W1AW THIS IS WB9ZZZ CLEAR AND LEAVING THE AIR

Listen to or copy a few QSOs on the air and soon you will have a good feeling for how things are said ham radio style!

INTRODUCTION TO PACKET RADIO

Besides CW and voice, there are other modes of communication that may be used by a radio amateur, including the newest mode called *packet radio*. Packet radio is so-called because the information that is intended for transmission is divided into packets before it is actually transmitted. The job of assembling and disassembling packets is performed by a *packet assembler-disassembler* or *PAD*. In the typical amateur packet radio installation, the PAD is supported by other circuits that are housed in a box called a *terminal node controller* or *TNC*. The TNC is connected to a source of information, typically a personal computer, and to radio equipment, typically an FM or SSB voice transceiver, that actually sends and receives the packets.

Briefly, the advantages of packet radio are that it provides error-free communications; that is, you can be assured that whatever you send will be received just as you sent it. Packet radio is also efficient because it transfers information more quickly than other amateur modes and many packet radio users can share the same frequency. Also, packet radio permits time-shifted communications. For example, you can send a message to another station at a time that is convenient to you and your message can be received later, at a time that is convenient to the other station.

Controlling the TNC

The TNC is controlled by a built-in microprocessor or microcomputer that is run by built-in software. Like most other computers, you control the TNC's computer by issuing commands to it. Typically, a TNC has over 100 commands to give you the versatility to use your TNC in any situation that arises. When the TNC sends the command prompt (cmd:) to your computer's display, it is ready to accept a command from you. To issue a command, you simply type the command at the command prompt, followed by a command modifier (if any) and a carriage return <CR>. (A command modifier is used with commands that configure your TNC's software for certain operating situations; for example, to set a TNC timing parameter, you would issue a timing command followed by the command modifier which represents the amount of time to be set.)

Station Identification

You never have to be concerned with station identification in the packet radio mode because your TNC automatically transmits your call sign whenever it is required during radio communications. Before the TNC can do that, though, you must program your call sign into the TNC's memory. This is accomplished by entering the "mycall" command at the command prompt followed by your call sign and a carriage return (for example, "mycall W1AW <CR>"). Having done that, you never have to reenter your call sign unless you wish to change it or you reset the TNC (thus, erasing your call sign from memory).

One radio amateur may operate multiple packet radio stations. To accommodate this, the TNC's software allows you to differentiate multiple operations by adding a number between 0 and 15 after the call sign (for example, W1AW-4). This number is called the *Secondary Station Identification* or *SSID*. Unless you specifically enter the SSID when you program your call sign with the mycall command, your SSID is assumed to be 0. For example, if you enter "mycall W1AW <CR>," the SSID is 0, however, if you enter "mycall W1AW-4 <CR>," then the SSID is 4.

HF versus VHF/UHF

In packet radio, the two most distinctive operating situations that you are likely to experience are operating in the HF and VHF/UHF bands. There are a number of important parameters that need to be set depending on your band of choice and, typically, all but one may be controlled from the keyboard of your computer. The one that cannot be controlled from your computer is the on-the-air operating speed or *data rate*. This parameter be set by selecting a multiposition switch on the rear panel of the TNC (refer to your TNC's manual for the correct switch settings).

For operation at 10 meters and below, your data rate should be set at 300 bauds and, for operation above 10 meters, your data rate should be set for 1200 bauds. (300 bauds below 28 MHz is the legal maximum data rate and is the data rate most commonly used at those frequencies. Above 50 MHz, data rates higher than 1200 bauds are permitted, but 1200 bauds is most commonly used.) The following table lists the other parameters that will optimize your TNC for HF or VHF/UHF operation.

command	*VHF/UHF*	*HF*
maxframe	4	1
paclen	128	64
dwait	16	0
digipeat	on	off

To program your TNC with these parameters, type the command at the command prompt, followed by the desired command modifier and a carriage return. For example, to set maxframe to 1, type: maxframe 1 <CR>.

Listening

As in any other mode of Amateur Radio communications, listen before you transmit in order to determine what is occurring on frequency. The most popular packet radio channels are in the 14.090 to 14.110 MHz or 28.100 to 28.105 MHz HF bands and in the 144.95 to 145.09 MHz or 221.01 to 221.09 MHz VHF bands. Tune your radio to an active packet radio channel and turn your TNC's listening function on by typing "monitor on <CR>" at the command prompt. This causes the TNC to display any packets that it receives in the following format: transmitting station call sign, >, receiving station call sign, : and contents of packet. For example, a packet displayed as:
WA1LOU>W1AW: Did you receive the propagation report?

This indicates that a packet has been sent by WA1LOU to W1AW. With this information, you know which stations are being received by your station and which stations you are probably capable of contacting.

Calling Another Station

To contact another station, simply enter the "connect" command followed by the call sign of the other station and a carriage return. For example, to contact W1AW, type, "connect W1AW <CR>" at the command prompt.

If the station you are attempting to contact is monitoring the same channel that you are transmitting on and is within the operating range of your station, a contact or *connection* will be made and your computer will display a message indicating the successful connection. For example, a successful connection with W1AW would result in the following message: ***CONNECTED to W1AW.

Calling CQ

To call CQ, first, type "unprotoco CQ <CR>" at the command prompt to properly configure your TNC, and type "convers <CR>" to transfer from the command mode to the converse mode. Finally, enter a carriage return to actually send a CQ. Each time you enter a carriage return, a CQ is transmitted and may be received by other stations on frequency who are within your station's operating range. If one of the stations receiving your CQ desires to contact you, that station may attempt to connect with you to initiate a conversation.

Exchanging Packets

Transmissions are exchanged so rapidly between the stations in packet radio that it is very similar to a face-to-face conversation. As a result, you do not need to send a "K" or "over" to indicate when you are through transmitting to let the other station know it is its turn to transmit. Both stations can practically transmit at the same time.

Ending a Connection

To end a connection, one of the stations must initiate a *disconnection*. This is accomplished by typing the Control-C (<CTRL-C>) control character to transfer to the command mode and, at the command prompt, typing "disconne <CR>." A successful disconnection is indicated by the "***DISCONNECTED" message.

Digipeaters and Network Nodes

On packet, it is permissible to use another station to extend your operating range. A simple example will demonstrate this capability. You wish to contact station X, but it is beyond your station's operating range. However, station Y is located between your station and station X and if it is within your operating range and within station's Y's operating range, you can use station Y to connect to station X. Station Y will act as a *digital repeater* or *digipeater* and repeat the packets sent between your station and station X.

Every TNC has the capability to act as a digipeater. Some stations, however, are purposely located at better radio locations to function as dedicated digipeaters; these are the ones you are likely to use more often. To use a digipeater, you simply append 'via' and the digipeater's call sign to the connect command. For example, to connect to W1AW using K1CE as a digipeater, type "connect W1AW via K1CE <CR>" at the command prompt. Once the connection is completed, you may transmit to the other station as if the digipeater was not there. Ending a connection is simply a matter of entering the disconnect command.

In addition to digipeaters, there are networks in place that can extend your operating range. There are different types of networks operating in various parts of the country, and user access to a network varies with the type in the user's area. In general, the user simply connects with a local network access point or *node* using the connect command and the node's call sign. Once connected, the user commands the node to con-

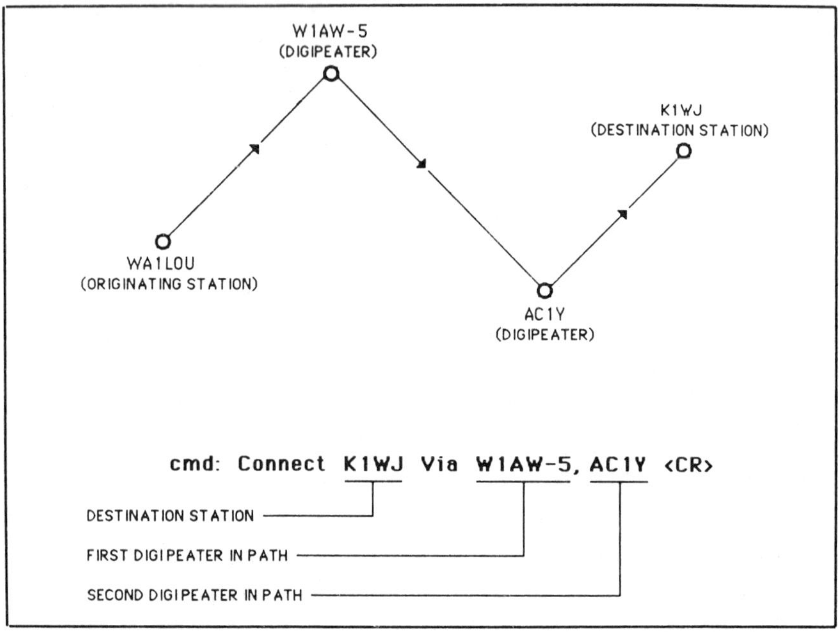

Networks of digipeaters are used to send messages over long distances.

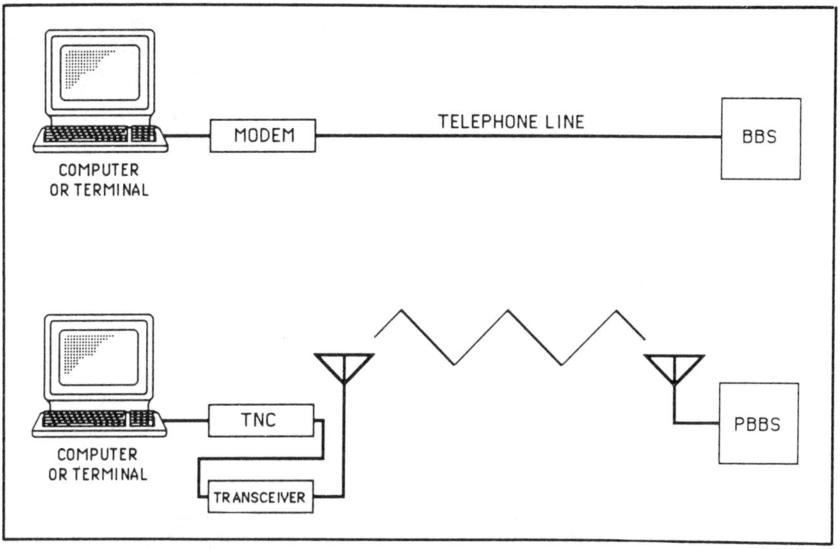

Packet-radio bulletin boards are similar to those used on the telephone line.

nect with a distant node and station. If a network is in operation in your area, you can probably obtain a description of its commands from the local node or from another user.

Time-Shifting Communications

Besides human operators, you may contact *packet radio bulletin board stations* or *PBBSs*, which are similar to the bulletin boards that are accessible with a computer and a modem via the telephone line. Like their phone line cousins, PBBSs contain messages that users send to other users, as well as general announcements, ARRL bulletins, computer programs, club newsletters and text files related to various Amateur Radio topics. Many PBBSs are part of a nationwide network that allows you to send messages to hams at other PBBSs throughout the country. You may also use this same network to send messages between nonhams (see "Packet-Based Traffic Handling" in the "National Traffic System" chapter of this book).

You connect to PBBSs in the same manner as you would any other stations (using the connect command and the PBBS's call sign). Once you are connected, the PBBS will send you a message welcoming you to the system. The typical PBBS contains numerous commands. When you connect to one initially, you should send it the H (for Help) command to obtain a description of its commands.

When you are finished using a PBBS, do not disconnect from it like you would a non-PBBS station, rather send it the B (for Bye) command. When the PBBS receives the B command, it will log you off and initiate a disconnection.

The Paper Chase: Logging, QSLs and Awards

LOGGING

Most of us keep a log which is a journal of our ham radio doings. Even though logs are no longer required by the FCC, it's important to keep a log for other reasons. Logs keep track of whom we talked to when and provide the information we need to confirm contacts with other stations in order to qualify for challenging operating awards. The basic information is simple: date, time, frequency, mode, signal reports (sent and received) and "notes." The *ARRL Log Book* is set up with preprinted columns for this information, and it contains basic operating information and country prefix lists as well.

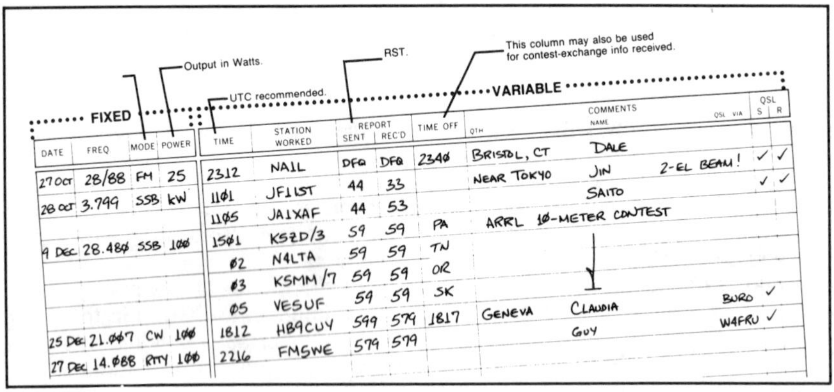

QSL CARDS

A QSL card serves as proof that you have contacted a particular station. Sending QSLs is a tradition in ham radio.

Whether you want the other station's QSL as a memento of an enjoyable QSO or for your collection of the cards necessary to apply for some ham radio operating award, it's a good idea to have your own QSL cards and know how to fill them out.

You can make your own cards (using post cards showing some local scene is popular), you can design your own card and have them printed locally, or you can check "Ham Ads" in *QST* where you'll find a couple dozen QSL printers listed—many of whom will send you free samples. Most hams have their call, name, location (city, state, ZIP or postal code, and country) printed directly on their cards. Other information such as county, grid square and equipment is often printed on cards. All the necessary information should appear on one side of the QSL. The rest of the information is normally written in the appropriate spaces on the card. Be neat. The standard size is 3½ inches high by 5½ inches long. Anything smaller cannot be mailed (except in an envelope); anything larger than 4 inches by 6 inches and they're tough to file. For most awards in general—and ARRL awards in particular—cards that have written-in information that is altered in any way or crossed out (and not initialed by the writer) won't be counted. The information that you fill in on your QSL to another station is basically an extract of the information recorded in the station log. Suppose WB9ZZZ is sending a QSL to W1AW. The information that WB9ZZZ hand prints on the card is shown on the sample card.

WISCONSIN

WB9ZZZ

QSO WITH	CONFIRMING QSO						
	DAY	MONTH	YEAR	UTC	MHz	RST	2 WAY
W1AW	6	1	89	0115	7	57	(SSB) CW

□ PSE QSL □ TNX QSL 73 Paul Smith
1200 Banta Lane
Donnelley, WI 53073

Practically all domestic cards are sent "direct," which means back and forth through the mail. For domestic QSLing it's a good idea to enclose an SASE (self-addressed, stamped envelope) if you are serious about getting a card in return.

Unless the other operator sends you his card first, you have to send the other station your card. Suppose you didn't get W1AW's address over the air, where do you send the card? *The Radio Amateur's Callbook* lists the addresses of stations by call sign. North American and International versions of the *Callbook* are advertised in *QST* every month.

QSL CARDS FOR CONTACTS OUTSIDE THE COUNTRY

So, how do you exchange QSL cards with DX stations in other countries? There are several options.

The ARRL DX QSL Bureau System

When you contact a DX station and ask for a QSL, the reply often will be that the other operator will "QSL via the bureau," or "via the buro" on CW. In each call area of the US and Canada, volunteers operate clearing houses for the distribution of DX QSL cards to amateurs in their area. To receive cards sent to you through the bureau, all you need to do is send an SASE to the bureau serving your area. Patience is a virtue for those waiting for cards coming via the bureau since it's not unusual for a card to be several months in transit. For further details and addresses of the bureaus send an SASE to ARRL HQ and ask for "Incoming QSL Bureau Information."

ARRL Membership Overseas QSL Service

The ARRL Membership Overseas QSL Service is the counterpart of the incoming bureau. This "outgoing" service allows ARRL members to send QSL cards to foreign countries at a minimum of cost and effort. An unlimited number of QSLs may be sent for distribution 12 times per year. The fee is just $2 per pound or portion thereof (150 cards average a pound.) The requirements are simple:

1) Presort your DX QSLs alphabetically by call sign prefix (A3, AP, C6, CE, F, G, GI, GM, I, JA, 4U1, and so on).

2) Enclose the address label from your current copy of *QST*. Family members may enclose their QSLs with those of the primary member. Include the appropriate fee with each individual's cards and indicate "family membership." Sightless members should indicate that the cards are from a "sightless member."

3) Enclose payment in the form of a check or money order (sending large amounts of cash through the mails is not suggested). Please do not send stamps.

You can get more information on the ARRL Membership Overseas QSL Service by sending an SASE to ARRL HQ.

QSL Managers

Some of the unsung heroes in Amateur Radio are QSL Managers. Often, DX stations will send copies of their logs to someone who will fill out their cards upon receipt of your card and SASE. If the DX station tells you to QSL via WB9ZZZ, it means that WB9ZZZ handles the DX station's QSL chores. It's also possible that the DX station is being operated by a visitor to the country who has a license there, but wants QSLs sent home. (The Regulatory Information Department at ARRL HQ has information on obtaining reciprocal operating permits for other countries.) There are QSL manager lists and the latest information is often found in the DX newsletters or the How's DX column in *QST*.

QSLing Directly to DX Stations

You can also send QSLs for foreign contacts "direct." Remember that the foreign ham cannot use US postage to mail a card back to you. The Post Office has this nifty thing called an IRC (International Reply Coupon) that the foreign ham can use to buy a stamp in his own country good for sending an envelope (which you've provided with your return address, but without postage) to you by surface mail. Want faster service? Check the chart in the *Callbook* that shows how many IRCs to enclose for return airmail service. You can sometimes buy IRCs at a discounted rate from DXers and QSL managers who would rather sell them to another ham than turn them over to the Post Office for US postage stamps.

Some foreign hams will ask for a "green stamp" (a US dollar bill) to cover postage, but beware: Possession of US currency in some countries is illegal. Sending a dollar to a station in such a country could get the DX operator in trouble. Never send a "green stamp" unless the DX operator requests it or you are absolutely sure that it will not cause problems for the DX station. If you're not sure, send IRCs. Although it's never wise to send currency through the mails, sometimes that's the only way to get a card from a rare station.

An alternative to IRCs and green stamps is to buy foreign postage stamps from a stamp dealer to put on the return envelope.

When All Else Fails

Remember that there are some hams who never QSL, and there is more than one ham who doesn't have a card from his first contact or his first DX QSO despite all efforts to obtain it. A ploy that sometimes works is to send a filled out card that just requires the signature of the station worked. Send an SASE of course.

It's said that QSLing is the final courtesy of the QSO and it's fun to exchange cards. Going through your QSL collection can bring back memories of "special" QSOs, and QSLs are worth collecting for operating awards too.

AWARDS

One of the most exciting aspects of Amateur Radio is awards chasing. There are hundreds of beautiful certificates and plaques available to hang on your ham shack wall. Some awards take little effort; others take years of operating skill and diligence to obtain. Here's just a small sample of the awards out there. These are available from ARRL. If you live in the US and possessions, Canada and Puerto Rico, you must be an ARRL or CRRL member to participate. In addition, official application forms are required for WAS, DXCC, VUCC and WAC. To obtain the forms, send a business size SASE to:

Special Requests
ARRL Headquarters
225 Main Street
Newington, CT 06111

RCC

Most awards hunters begin with the *Rag Chewers' Club* certificate. This award encourages friendly, meaningful contacts rather than just

exchanging signal reports in "hello-goodbye" contacts. Chew the rag for at least a half hour, and report the details of the QSO to ARRL HQ. Send along a business size envelope, and your Rag Chewers' Club Certificate will be enroute to you by return mail.

WAS

Talk to ("work" in ham radio lingo) other hams in 50 states, get their QSL cards, have these cards checked either at ARRL HQ or a Special Service Club HF Awards Manager, and you'll qualify for the *Worked All States* award. Besides the basic certificate for any combination of bands/modes, specialty certificates are available for a variety of different bands and modes such as satellite, 160 meters, and each VHF band. Endorsement stickers are available for SSB, CW, Novice, moonbounce, packet, QRP (low power) and for each single band. The *5-Band WAS* award (a certificate and plaque) is available for working all states on five amateur bands (except (1.8/10/18/24 MHz). Contact ARRL HQ for the nearest ARRL HF Awards Manager who can check your cards.

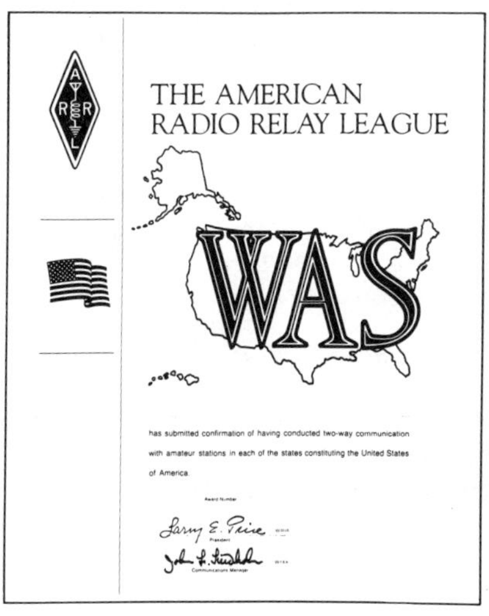

WAC

The *Worked All Continents* award is sponsored by the International Amateur Radio Union (IARU) for working and confirming all six continents (Africa, Asia, Europe, North America, South America and Oceania) on a variety of different bands and modes. A *5-Band WAC* certificate and a 6-band sticker are also available. For US amateurs, cards are checked at ARRL HQ on behalf of the IARU.

DXCC

DX is Amateur Radio shorthand for long distance. It's also universally understood by hams that a "DX" station is one in a foreign country. If you can work and confirm at least 100 countries on the ARRL DXCC Countries List, you can qualify for the premier operating award in all of Amateur Radio: The *DX Century Club.* All cards are checked at the DXCC Desk at ARRL HQ. After working the first 100 countries, you can qualify for endorsements in certain increments on up through the coveted DXCC Honor Roll for those at or near working them all. The basic certificate is available in these

categories: mixed modes, phone, CW, RTTY, 160 meters, 80 meters, 40 meters, 10 meters, 6 meters, 2 meters and satellite. The *5-Band DXCC* is available as a certificate or plaque for working 100 countries on each of five bands (except 10 MHz) and endorseable on 1.8, 18, 24 and 50 MHz.

VUCC

Here's an award for the VHF/UHF operator. It's issued for working and confirming contacts in 2° × 1° grid square locators on bands from 50 MHz to 47 GHz. Individual certificates are issued per band to those meeting initial qualifying levels, and there is an extensive endorsement program. Contact ARRL HQ for the nearest ARRL VHF Awards Manager who can check your cards.

Code Proficiency

This award is based not on collecting QSL cards, but on your ability to copy Morse code. Copy one of W1AW's twice-monthly qualifying runs (sent at speeds from 10 to 40 WPM—see the W1AW schedule and Contest Corral column in *QST*) and submit one minute of solid copy to ARRL HQ. Your submission will be checked directly against the official W1AW text, and if you pass, you'll get your initial Code Proficiency certificate. Endorsement stickers are issued for speeds up to 40 WPM.

Other Certificates and Awards

Old Timer's Club. The *OTC* certificate is available if you have been licensed for twenty years or more. Drop a note to HQ with the date of your first license and your call (both then and now) plus a business-size SASE, and HQ will send your Old Timer's Club certificate promptly.

A-1 Operator Club. Only the best operators can qualify as an *A-1 Op.* You can't apply; you must be recommended independently by two amateurs who already belong to the A-1 Operators Club. Fine-tune your operating skills so they meet the highest operating standards, and some day you will be pleasantly surprised when the Postal Service delivers an A-1 Operator Club certificate.

Extra Class Certificate. When the FCC stopped giving out certificates for attaining the highest grade of Amateur Radio license, ARRL took on the job. Send a photocopy of your license and a fee of $3 to ARRL HQ, and this attractive certificate will be mailed back to you.

In addition to complete details on the ARRL awards program, *The ARRL Operating Manual* also lists hundreds of awards sponsored by other national and international organizations. For instance, there are awards for working all counties and parishes in the US. The Worked All Zones awards from *CQ* Magazine are popular. All these awards attest to your skill and achievement as a radio amateur.

Operating Activities

RADIO CLUBS

What better place to discuss your operating activities and receive advice (hams are good at giving each other advice) than at a radio club meeting. Some even have "Beam Teams" or "Quad Squads" to help members with their antenna projects.

Most clubs are geographical in nature and general interest in scope—The Podunk Hollow Radio Club for instance. There are also specialized clubs for traffic handlers, VHFers, contesters, DXers, FM repeater enthusiasts and training. Many of these have met the requirements to be ARRL Special Service Clubs. For a list of all affiliated clubs in your area, send an SASE to:

Field Services Department
ARRL HQ
225 Main Street
Newington, CT 06111

DXING

We've touched briefly on the art of getting QSL cards from DX stations and mentioned the DX Century Club (DXCC) award. One of the most interesting and exciting facets of Amateur Radio is DXing—talking to hams in foreign lands. A half-dozen books have been written on the subject of DXing, and even more exist covering radio-wave propagation—how signals get from here to there. *The ARRL Operating Manual* devotes two chapters to DXing, and there are supporting chapters such as the one on Antenna Orientation. Two good books on the subject are *The Complete DX'er* and *The DXCC Companion,* which are available from ARRL; both cover the subject well.

Finding DX

You don't have to be an expert DXer to have a lot of fun. Even with

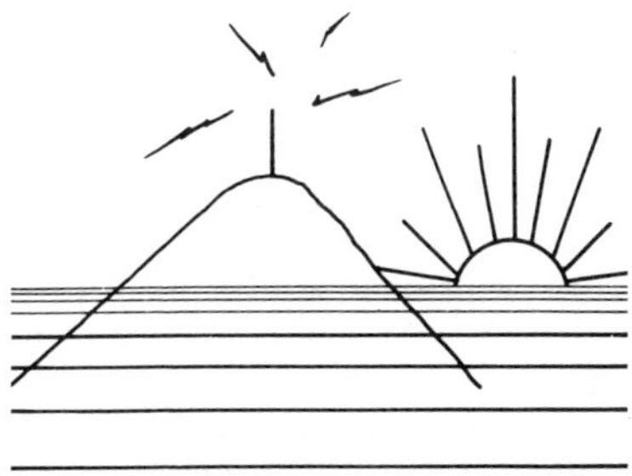

a modest station you can expect to work a whole lot of DX, especially on 10 meters. The key is, of course, to listen, listen, listen! Rare DX stations often listen on a frequency a little above or below their transmit frequencies, and announce where they're listening. You'll have to make sure you are transmitting on the announced frequency in order to be heard, and be able to hear the DX station on his transmit frequency when he comes back to you. This is known as *split-frequency* operation and can be accomplished with a transceiver with two VFOs, or frequency memories. If the station is listening close enough to his transmit frequency, you can use your transceiver RIT (receiver incremental tuning), which is a sort of fine tuning control.

Mr Sun

Solar radiation ionizes various layers of the ionosphere causing radio signals to "skip" or be reflected back to earth. Here are some very basic propagation characteristics of some of our popular bands.

160 and 80 meters. These are regarded as nighttime DX bands, although some of the best propagation occurs just before or after sunset or sunrise. Both bands suffer from static in the summer.

40 meters. Working DX on 40

meters is a lot easier than on 160 or 80. Even though it's regarded as a nighttime band for DX it holds some nice secrets like mid-afternoon DX during winter months.

20 meters. Twenty is regarded as the prime DX band. When sunspot numbers are high it can provide worldwide communication 24 hours a day. When conditions aren't so good, it will open at sunrise and you'll be able to work stations to the East. At sunset before the band closes, you'll be able to work stations to the West. Remember that the earth is round and radio waves usually travel over the shortest path, but more than 80 or 40, 20 meters also supports *long-path* propagation in the direction opposite of the short-path direction.

15 meters. This band is a lot of fun. Fifteen usually opens after twenty and closes before twenty does. Sometimes fifteen can stay open all night if conditions are really good. It might open just briefly and support only north-south communication when conditions are poor.

10 meters. Opens after 15 and closes sooner. There's either feast or famine on this band depending on how active the sun is. Just like on 15, you can usually find a few South American stations to work even when conditions are poor. When the band is open at the peak of the sunspot cycle, worldwide communication can be accomplished with a few milliwatts of power. During certain times of the year, particularly during the summer months, sporadic E propagation can provide contacts up to 1500 miles (twice that if "double-hop" conditions occur) on ten.

30, 17 and 12 meters. Awards credits for contacts on these newer bands vary from band to band and award to award. Check the official rules for the latest status.

CONTESTS

The first contests were operating periods set aside specifically to attempt transoceanic communication between North America and Europe just after World War I. Through this coordinated effort, many successful contacts were made "across the pond." We hams had done what the experts said *couldn't be done* on such high frequencies. This activity evolved into what is today the ARRL International DX Contest—held in February and March. While the rules have changed over the years, the object is still to contact as many stations in as many areas of the globe as possible during the contest period. What a quick way to increase your total of DXCC countries worked!

ARRL Sweepstakes

The US and Canada are divided into 77 sections (69 in the US, the rest in Canada). In the ARRL November Sweepstakes Contest, it's not uncommon for stations to work all sections or at least talk to stations in every state and province during the 24-hour operating period.

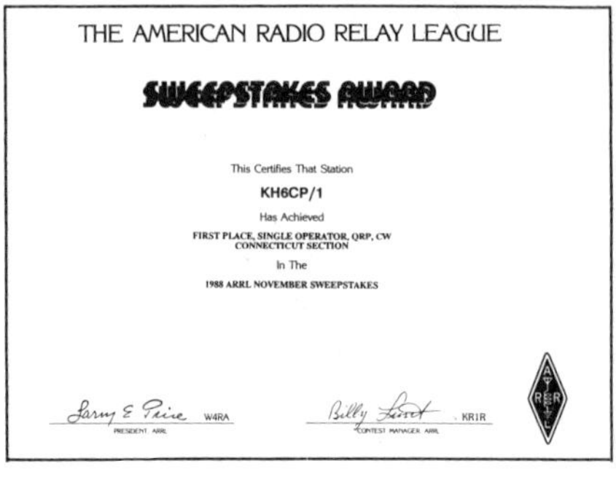
THE AMERICAN RADIO RELAY LEAGUE

SWEEPSTAKES AWARD

This Certifies That Station

KH6CP/1

Has Achieved

FIRST PLACE, SINGLE OPERATOR, QRP, CW
CONNECTICUT SECTION

In The

1988 ARRL NOVEMBER SWEEPSTAKES

W4RA — PRESIDENT ARRL

KR1R — CONTEST MANAGER ARRL

Novice Roundup

There's a week-long Novice Roundup for beginners where Novices and Technicians try to work as many stations as possible—in as many ARRL sections as possible—in the 30 hours total operating time allowed. This event, usually held the week that straddles the end of January and the beginning of February, is a great way for you to get your feet wet.

THE AMERICAN RADIO RELAY LEAGUE, INC

HEADQUARTERS, NEWINGTON, CONNECTICUT

NOVICE ROUNDUP AWARD

This certifies that Amateur Radio station N1FOZ/T has participated in the 1989 ARRL Novice Roundup contest, as documented in QST.

W4RA PRESIDENT, ARRL

KR1R CONTEST MANAGER, ARRL

Getting into the Fray

If you hear stations operating in a contest, it's usually pretty easy to pick up on what information is being exchanged. Then, you can join in. But what contest is it? Check the Contest Corral column in *QST* for the particulars. Don't ask one of the participants what is going on. Serious contesters are running a radio marathon! Don't upset the contester's rhythm even by asking "Dragnet-style" questions. (You wouldn't stop one of the runners in a race to ask who is sponsoring the event, how far it is and how to enter.)

Win, No Matter What!

Set some personal goals that are attainable. If you're a beginner, you're probably not going to be able to work stations very fast. If you worked seven stations in the last hour, try to work ten the next hour. Be brief. Send the required information only *once.* Wasted words mean wasted time, and avid contesters keep an eye on the clock. The experienced contesters will appreciate your "snappy" operating.

Keep a matrix of stations contacted. This is called a *dupe sheet;* by checking to make sure you haven't worked a station before, you'll prevent duplicate contacts or "dupes." Dupes waste the time of both stations and they must be crossed out of the log before it is submitted for checking. In most contests, you can work a station once per band. The ARRL Sweepstakes is an exception: You can work a station only once during the entire contest. Printed dupe sheets are available free from ARRL HQ for an SASE.

When you first start out, it is preferable to concentrate on hunting and pouncing, that is, tuning for stations calling "CQ Contest" and responding to them. As you progress, you will eventually be able to run strings of stations by calling CQ yourself, but even contesters with high rates have to hunt and pounce to find stations located in rare multipliers.

Suggested Standard File Format for Submission of ARRL Contest Log Data

All files must be in standard ASCII format.

All files must be on an MS-DOS formatted disk, 3.5 inch (720-kB or 1.44-MB) or 5.25 inch (360-kB or 1.2-MB).

A diskette label should clearly indicate the call used, contest name, entry class, and date of the contest.

The log file must consist of one line of data per QSO, without headers, footers, page breaks or other non-ASCII characters.

All QSO data must appear in each line, aligned by columns, and must include the following:

- band: wavelength for HF and frequency for VHF.
- mode: designator such as CW, PH, etc.
- date: in MM/DD/YY or DD/MM/YY format.
- time: 4-digit UTC without colons.
- station worked.
- complete exchange sent.
- complete exchange received.
- indication of multipliers: for example an "*", section/prefix/zone or etc.
- points claimed: ALL unclaimed QSOs must be "0" points.
- Multioperator, two-transmitter category log in the ARRL International DX Contest must indicate which transmitter made which QSOs.

The log data file name shall consist of the call sign and the extension ".LOG".

A paper summary sheet is required of all logs, either an official ARRL summary sheet or a close facsimile with a signed contest participation disclaimer.

All diskettes become the property of the ARRL and are not returnable.

Scoring

In most contests, the score is determined by multiplying the number of contacts, each with a certain point value, by the number of geographical areas (such as a state) that you've contacted. During the ARRL Sweepstakes there may be hundreds of stations on from Ohio, but in the sparsely populated Yukon Territory there may be only one station on. That's what is meant by a rare multiplier!

When logs are submitted for scoring they can be—and often are—checked against one another to verify the accuracy of log

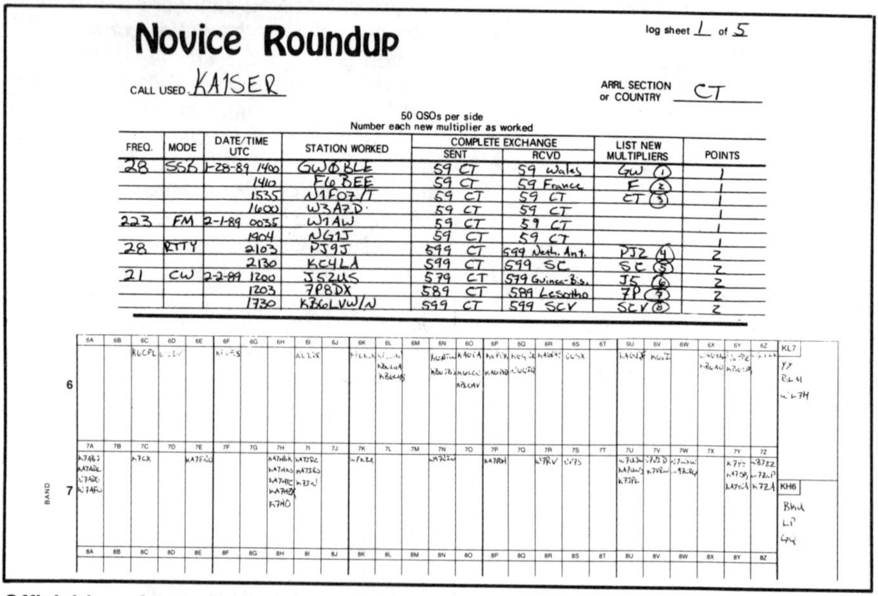

Novice Roundup

log sheet 1 of 5

CALL USED KA1SER

ARRL SECTION or COUNTRY CT

50 QSOs per side
Number each new multiplier as worked

FREQ.	MODE	DATE/TIME UTC	STATION WORKED	COMPLETE EXCHANGE SENT	COMPLETE EXCHANGE RCVD	LIST NEW MULTIPLIERS	POINTS
28	SSB	1-28-89 1400	GW0BLE	59 CT	59 Wales	GW (1)	1
		1410	F6BEE	59 CT	59 France	F (2)	1
		1535	N1FOZ/T	59 CT	59 CT	CT (3)	1
		1600	W3AZD	59 CT	59 CT		1
223	FM	2-1-89 0035	W1AW	59 CT	59 CT		1
		1904	NG1J	59 CT	59 CT		1
28	RTTY	2103	PJ9J	599 CT	599 Neth. Ant.	PJ2 (4)	2
		2130	KC4LA	599 CT	599 SC	SC (5)	2
21	CW	2-2-89 1200	J52US	579 CT	579 Guinea-Bis.	J5 (6)	2
		1203	7P8DX	589 CT	589 Lesotho	7P (7)	2
		1730	KB6LVW/N	599 CT	599 SCV	SCV (8)	2

Official log sheets (top) are usually available from contest sponsors. "Dupe sheets" (bottom) are used to tell if the same station has been contacted more than once.

information. No one is perfect, but excessive duplicate contacts, miscopied calls and QSO information, and nonexistent QSOs can result in a station being listed in the disqualification section of the score results. 'Nuff said.

Standard Contest Guidelines

1) Where possible, use preprinted log sheets, dupe sheets and summary sheets which are usually available from the organization that sponsors the contest.

2) Be sure your log details the date, time, band, call sign and complete exchange sent and received for each QSO claimed for contest credit.

3) The summary sheet should indicate your score (clearly show how you figured it) and a declaration that you followed contest rules and FCC regulations (or the regulations of the licensing authority in your country). Your name, call sign and complete address should be typed or printed in block letters.

4) Crossband, crossmode and repeater contacts are never permitted.

5) Check your log carefully for duplicate contacts. If you made more than 200 contacts, dupe sheets should be included with your entry.

6) Your log may be considered a checklog or be disqualified if it is incomplete or if too many errors are detected by the contest committee.

7) Avoid standard net frequencies, listen before you transmit and ask if the frequency is in use (QRL? on CW) before transmitting.

8) ARRL contests generally offer awards to top scorers in each ARRL section. International contests generally offer awards to top scorers from each US call area and each country. Domestic contests such as state QSO parties usually offer awards to the top scorers in each state and province.

For more information on contesting check out *The ARRL Operating Manual.* Also, avid contesters should subscribe to the *National Contest Journal,* a magazine about competitive Amateur Radio that is

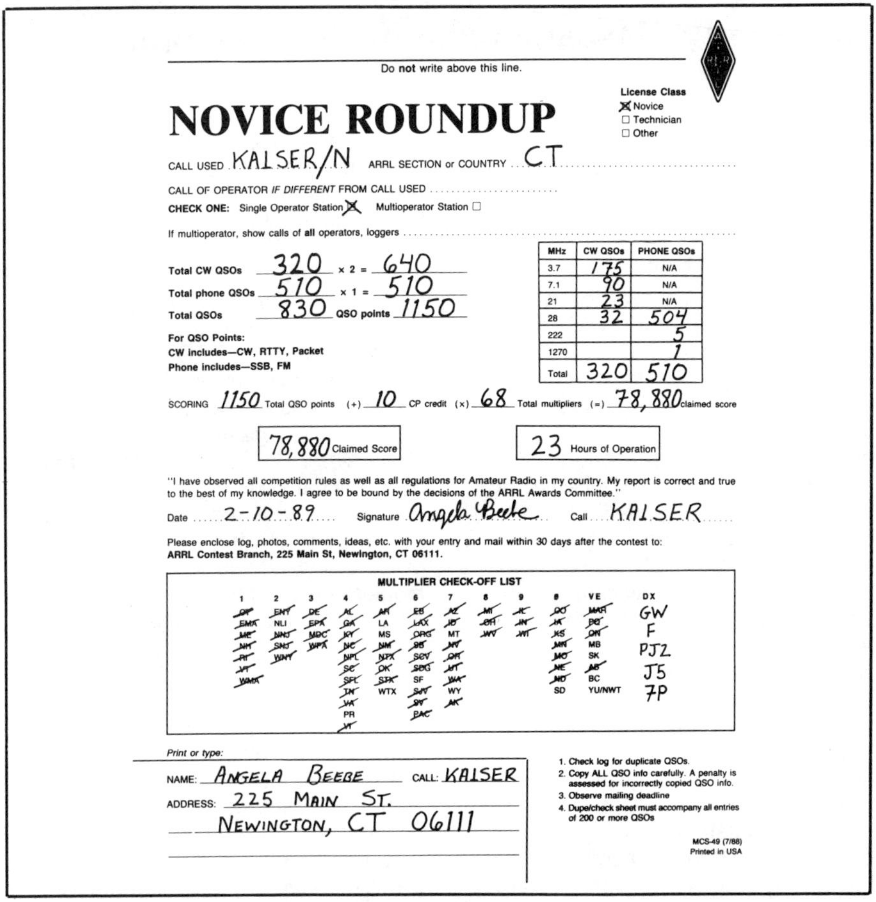

Do **not** write above this line.

NOVICE ROUNDUP

License Class
☒ Novice
☐ Technician
☐ Other

CALL USED KA1SER/N ARRL SECTION or COUNTRY CT

CALL OF OPERATOR *IF DIFFERENT* FROM CALL USED

CHECK ONE: Single Operator Station ☒ Multioperator Station ☐

If multioperator, show calls of **all** operators, loggers

Total CW QSOs 320 × 2 = 640
Total phone QSOs 510 × 1 = 510
Total QSOs 830 **QSO points** 1150

For QSO Points:
CW includes—CW, RTTY, Packet
Phone includes—SSB, FM

MHz	CW QSOs	PHONE QSOs
3.7	175	N/A
7.1	90	N/A
21	23	N/A
28	32	504
222		5
1270		1
Total	320	510

SCORING 1150 Total QSO points (+) 10 CP credit (×) 68 Total multipliers (=) 78,880 claimed score

78,880 Claimed Score

23 Hours of Operation

"I have observed all competition rules as well as all regulations for Amateur Radio in my country. My report is correct and true to the best of my knowledge. I agree to be bound by the decisions of the ARRL Awards Committee."

Date 2-10-89 Signature Angela Beebe Call KA1SER

Please enclose log, photos, comments, ideas, etc. with your entry and mail within 30 days after the contest to:
ARRL Contest Branch, 225 Main St, Newington, CT 06111.

MULTIPLIER CHECK-OFF LIST

1	2	3	4	5	6	7	8	9	ø	VE	DX
CT	ENY	DE	AL	AR	EB	AZ	MI	IL	CO	MAR	GW
EMA	NLI	EPA	GA	LA	LAX	ID	OH	IN	IA	PQ	F
ME	NNJ	MDC	KY	MS	ORG	MT	WV	WI	KS	ON	PJ2
NH	SNJ	WPA	NC	NM	SB	NV			MN	MB	J5
RI	WNY		NFL	NTX	SCV	OR			MO	SK	7P
VT			SC	OK	SDG	UT			NE	AB	
WMA			SFL	STX	SF	WA			ND	BC	
			TN	WTX	SJV	WY			SD	YU/NWT	
			VA		SV	AK					
			PR		PAC						
			VI								

Print or type:

NAME: ANGELA BEEBE CALL: KA1SER
ADDRESS: 225 MAIN ST.
NEWINGTON, CT 06111

1. Check log for duplicate QSOs.
2. Copy ALL QSO info carefully. A penalty is assessed for incorrectly copied QSO info.
3. Observe mailing deadline
4. Dupe/check sheet must accompany all entries of 200 or more QSOs

MCS-49 (7/88)
Printed in USA

A summary sheet shows the contest sponsor how you figured your score.

published by the ARRL every other month. Contesters get to read about station setup, optimizing equipment, antennas, score information and articles about contest operation.

ROUNDTABLES AND NETS

Most hams start out by getting on the air and making contacts at random by calling or answering CQs. This makes for fun and enjoyment. You might, after a period of time, start talking to the same station on the same frequency at the same time every day (or week) on a regular schedule. This organized activity is known as a "sked." Suppose you ask a third and maybe a fourth station to join you on a regular basis. Then you have what is called a roundtable.

A formal roundtable with a "master of ceremonies" which is organized for a specific purpose is called a *net.* The "MC" is called the *net control station* (*NCS* on CW). There are all sorts of nets: nets organized just for ragchewing, emergency nets, weather nets, DX nets, nets where contesters gather and slow-speed CW nets for beginners, for example. But that's not all: National Traffic System (NTS) Nets and Packet Radio Bulletin Boards are covered in a separate section later.

With all these nets, why not put out a directory of them? Well, have we got a deal for you! Most nets that have a purpose other than for ragchewing are listed in the *ARRL Net Directory.* Listings show the name of the net, frequency, time, days of operation and purpose. It's available for $1.00 from ARRL HQ.

Third-Party Messages and Traffic Handling

The handling of message traffic was the basis for the formation of the ARRL and this is the principal operating activity for a sizable segment of radio amateurs.

THIRD-PARTY TRAFFIC

A third-party message doesn't have to be formal at all, and can be as simple as saying "Tell my wife I'll be late getting home for dinner." There are three types of third-party traffic. The first deals with repeater autopatch and phone patches. The second is the direct participation by an interested third party in Amateur Radio communication. The third is the popular activity of relaying messages. Almost all traffic handled by US amateurs is originated (begun) and delivered within the US. But, when messages have to cross the borders of countries, all sorts of international regulations come into play. *The FCC Rule Book*, which is published by ARRL, devotes considerable space to FCC and international regulations covering third-party traffic. Here are some guidelines as to who, what, when, where and how messages can be sent.

When to Say "No" and When to Say "Yes"

Several dozen countries have told our government that they have no official objection if amateurs in their countries receive unimportant, personal nonbusiness messages. These are the countries with whom the US has third-party agreements. Check the "Mini Directory" that appears each month in *QST* for a recent issue that lists the countries with whom we have third-party traffic agreements or check *The FCC Rule Book*.

We should define what unimportant means. In most countries, the telephone company is a government monopoly, and the government does not want its revenues compromised by anyone else—including hams—who can relay messages. The first test is that you wouldn't pay to telephone someone to give them the message.

You can't be paid for handling traffic no matter where it's going, That's the "without pecuniary interest" part of the FCC's definition of the Amateur Radio Service. Business communications are also prohibited by our rules. Again, you should check *The FCC Rule Book* for examples of what you can and cannot do.

Emergency communications must relate directly to the *immediate* safety of life of individuals or to the *immediate* protection of property. In an emergency, the FCC does not prohibit US amateurs from handling traffic relating to the saving of lives and property.

The FCC and the third-party traffic rules are concerned with the *borders* of the countries involved. Say your station is operating in the US. Your concern should be that we have a third-party agreement with the country of the station you first contact to relay the message, and not the destination country. The same is true for messages coming to you. An agreement must exist between the US and the country of the station that you've contacted that is relaying the message, not the country where the message was originated.

The Autopatch

An autopatch is a device that allows repeater users to make telephone calls through the repeater. Tone pads are often mounted on the front of a portable transceiver or on the microphone. The tone pad generates telephone company tones to access and dial through the system. If you are new to a repeater, you'll probably have to ask for information on how to access and use the autopatch. Some autopatches have autodialers so that you only have to key in a few digits for commonly used numbers. Most of these are set up to access the phone numbers of emergency services like 9-1-1. (If you do access 9-1-1, remember chances are you are talking to a dispatcher in the town where the repeater is located, and not necessarily in the area where you are transmitting.)

An autopatch shouldn't be used where regular telephone service is available. Is the call necessary? If it is an emergency situation, sure. In fact many repeater groups limit the use of their autopatch to emergencies only. Again, business communications are strictly prohibited. Sorry, but you can't call the local pizza palace to place a take-out order even if it's "negative on the anchovies!"

Phone Patches

Hams have been using phone patches on our HF bands for years. The most common use is to allow a fellow ham in a country with whom we have a third-party agreement to talk through your station with his or her family in the states. With your radio hooked up to your telephone, you call the family using your regular telephone (or tone pad) and they can transmit and hear through your station. You also have to be able to hear the conversation, and have control of your station.

Third-party conversations must not be retransmitted until the people involved understand the nature of Amateur Radio. Besides the fact that they should say "over" when they want the other station to talk, they should be warned about business communications and that the information they convey can be heard by the public. It's up to you to make sure that there are no violations of regulations while carrying out a phone patch. Because phone patches (and autopatches at the repeater site) are wired to the telephone lines, the interconnections must be in accordance with telephone company rules (called tariffs).

ARRL has a policy to safeguard the prerogative of amateurs to interconnect their stations, including repeaters, to telephone lines. An important element of this defense is encouraging amateurs to maintain a high standard of legal and ethical conduct in their patching activities.

Letting Third Parties Talk From Your Station

Anyone, amateur or nonamateur, may participate in Amateur Radio communication as a third party. Remember that your guest can only talk to individuals in countries with whom we have third-party agreements. A control operator is required to be present to supervise and control the operation.

Who is the *control operator*? The licensee and the control operator of the station share responsibility. The station licensee is the person whose name appears on the station license. The control operator is the person who sits at the controls (control point) to perform the immediate operation of the station. Usually, the station licensee and the control operator are the same person—you normally operate your own rig. But occasionally you will invite a ham friend over to your shack to, say, help share operating duties in a contest. When your friend is controlling the station, he or she is the control operator.

Suppose you have a guest who has a Novice class license and you happen to be a General class licensee. The Novice can operate your equipment only to the extent of his or her Novice class privileges. For example, he or she can operate your station with 200 watts of power between 28.3 and 28.5 MHz. You don't even have to be there. But the second your kilowatt linear amplifier goes on or operation takes place on, say, 28.6 MHz, our Novice friend becomes a participant and is no longer a control operator. Now third-party rules apply and you (the General class licensee) must be there as control operator. It's okay for the third party to identify your station, *but* a third-party participant may only communicate with US stations and those countries with whom we have third-party agreements.

ORIGINATING AND RELAYING MESSAGES

All ready to get into the traffic handling game? *The ARRL Operating Manual* has an extensive and complete chapter on traffic handling. Messages handled by radio are—no surprise—called *radiograms*.

An amateur message is first originated, then relayed and finally delivered. To make the traffic flow smoothly, there are certain forms and procedures for message handling. The station that initiates the message is called the "originator." You can originate a message on behalf of another individual whether such an individual is licensed or not. It's up to the originating station to see that the message is in its proper format.

Each message originated and handled should contain the following parts sent in the order given:

1) Number
2) Precedence
3) Handling Instructions*
4) Station of origin
5) Check
6) Place of Origin
7) Time Filed*
8) Date
9) Address
10) Text
11) Signature

*Optional

Yikes! That seems complicated. Like everything else, you'll get used to the message form, and here is what all those strange words mean.

Items 1 through 9 make up the *Preamble* of the message.

1) You assign a sequential *number* to each message you originate. Your first message is number 1, and so on. You can set up a number sheet and after the message is sent, you can list the date and the station you first relayed the message to.

2) There are four *precedences* that tell the stations relaying the message how important the message is. Most messages take the *R* or *routine* precedence. Others are *EMERGENCY* which is always spelled out and meets the definition mentioned previously. The *W* or *Welfare* precedence is used on messages asking about the health and welfare of an individual in a disaster area, or on messages originated from within a disaster disaster area saying all is well. If all is not well, then the *P* or *Priority* precedence is used to indicate the message is of high importance. Those messages with a time limit, official messages not covered by the emergency category, press dispatches, emergency related traffic not of the utmost urgency, or notices of death or injury in a disaster area (personal or official) qualify for the priority precedence.

3) *Handling instructions* are optional cues to handle a message in a specific way. For example, HXG tells us to cancel delivery if it requires a toll call or mail delivery, but to let us know by a return message that the message has been canceled. (This return message is called a *service message*.)

4) The *station of origin* is the call of the station that originated the message and is never changed.

5) The *check* (CK on CW) is the number of words of *text* in the message. Remember: The check is a count of the number of words in the text of the message (item 10) *only*. This excludes everything before the text and the signature. The receiving station will count the number of words he has written down in the text, and that number should equal the number in the check that you sent.

6) *Place of origin* is usually the city and state of the originating station. It can also be the location of the third party wishing to initiate a message through the originating station.

7) The *filing time* is optional and is the time in UTC the message was filed at the originating station.

8) The *date* is the month and day (year is not needed) that the message was filed at the orginating station.

9) The *address* is the name, street and number, city and state of the party to whom the message is being sent. The telephone number of the person who is to get the message should also be part of the address.

10) The *text* is the message itself. It is normally hand-written 5 words per line (10 if typed). That makes counting words for the check a snap!

11) Finally, the *signature*. Remember that words like "sincerely" or other complementary closings are part of the text.

Of X-Rays, $\overline{\text{AA}}$s, Breaks, $\overline{\text{BT}}$s, $\overline{\text{AR}}$s, Bs, Ns, More and No More

The term "X-ray" is often used for the period. On CW, punctuation marks are spelled out, though the X-rays are just sent as an X.

On CW, the prosign $\overline{\text{AA}}$ (sent as you would $\overline{\text{SK}}$ or $\overline{\text{AR}}$) is used between *each* line of the address. The prosign $\overline{\text{BT}}$ is used to separate the Preamble (items 1 through 9) and the text. Another $\overline{\text{BT}}$ is used to separate the text from the signature. After the signature is sent, you say "end of message" on voice and send $\overline{\text{AR}}$ on CW.

On voice, if you have more messages you say "more to follow." On CW, you send B. If you don't have any more messages to send, you say "no more" or send an N on CW.

Transmitting the Message

We're finally ready to transmit the message. ARRL message forms, which you can purchase from HQ, make it easy to remember all of the information that goes into sending a message. All you do is fill in the blanks! Good news. What goes for transmitting a message that you've originated and one that you've received and are going to relay on

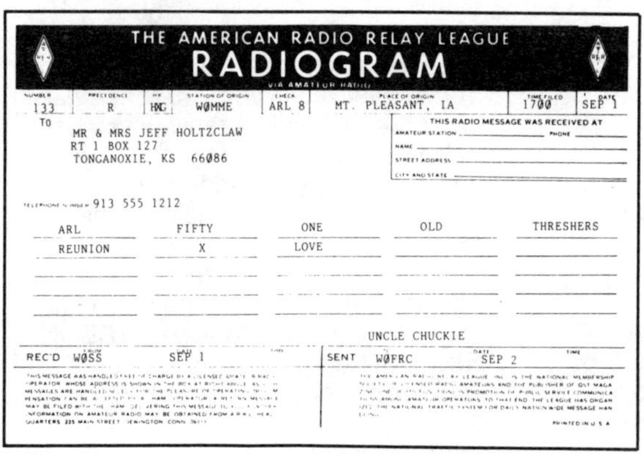
THE AMERICAN RADIO RELAY LEAGUE

RADIOGRAM

VIA AMATEUR RADIO

NUMBER	PRECEDENCE	HX	STATION OF ORIGIN	CHECK	PLACE OF ORIGIN	TIME FILED	DATE
133	R	HXG	WØMME	ARL 8	MT. PLEASANT, IA	17ØØ	SEP 1

TO
MR & MRS JEFF HOLTZCLAW
RT 1 BOX 127
TONGANOXIE, KS 66Ø86

THIS RADIO MESSAGE WAS RECEIVED AT

TELEPHONE NUMBER 913 555 1212

ARL	FIFTY	ONE	OLD	THRESHERS
REUNION	X	LOVE		

UNCLE CHUCKIE

REC'D WØSS SEP 1 — SENT WØFRC SEP 2

ARRL message form. The letters ARL inserted in the preamble in the check and text before the spelled out numbers represent standard text from the list of Numbered Radiograms which appears in the *ARRL Log Book*, or is available from ARRL HQ. When this message is delivered it would be read "Greetings by Amateur Radio. This message is sent as a free public service by ham radio operators here at the Old Threshers Reunion. Am having a wonderful time."

are the same. The next page shows the message we are going to send.

On voice it will go like this: "Copy number one, routine, WB9ZZZ, ten, Chicago, Illinois, 2125 Zulu, April 12. Miss Michelle Smith, one nine zero eight Moon Street Northeast, Albuquerque, New Mexico, eight seven one one two, telephone two nine eight six four zero eight. Break. Mother and Dad arrived home safely Sunday afternoon X-ray love. Break Bob. End of message, no more."

On CW: NR 1 R WB9ZZZ 10 CHICAGO IL 2125Z APR 12 MISS MICHELLE SMITH AA 1908 MOON STREET NORTHEAST AA ALBUQUERQUE NM 87112 AA 298 6408 BT MOTHER AND DAD ARRIVED HOME SAFELY SUNDAY AFTERNOON X LOVE BT BOB AR N.

Digital modes: Later on we have a whole section on packet-based traffic handling.

Oops—Missed Some

If the receiving station didn't receive all of the message, he will ask for fills. Say the street was garbled. The receiving station will simply ask "street?" Perhaps the last part of the text wasn't copied. The receiving station will ask you for "all after arrived." The receiving station missed everything after the word "arrived," so you will have to repeat the part of the message after "arrived." Don't repeat the whole message!

Common abbreviations used on CW for fills are AA (all after), AB (all before), WA (word after), WB (word before), or BN (between). Digital communication modes have *handshaking* between the transmitting station and the receiving station to ensure that all of the message is received.

Receiving a Message

We've already covered "fills." On phone, most stations let their VOX drop so that they can hear between sentences so if you miss something you can easily let them know. On CW, most experienced ops use break-in, which lets the transmitting station hear between words, if not between letters! Even without break-in, the transmitting station can be broken when he pauses between AAs or BTs.

Relaying a Message

You know how to receive a message and how to transmit one. Relaying is a simple combination of both. You send the message just as it was received. The only thing you can change is to add a corrected check *only if* there is a problem with the word count. You send the original check corrected to the word count you have. That's it! Well. . . not quite.

We haven't told you how to find a station to send the message to. Now you might get on the air and call "CQ Albuquerque" and you might find a station who will deliver the message for you. But, there's a better way. Get the message into the ARRL National Traffic System. NTS is covered in detail in the next section, and that's how most formal radiogram-type messages are relayed.

Delivering the Goods

Suppose you pick up some traffic on a net that's for someone in your immediate area. Delivery is the object of all message handling and a message should be delivered as soon as it is within nontoll telephone range. A good way to start out (so you don't scare the person half to death to whom you are delivering the message) is to say "This is Amateur Radio operator Bob Smith, and I've received a greeting message from a relative of yours." Speak clearly and offer to handle a reply. If you can't deliver the message for some reason, send a service message back to the originating station, explaining that the message can't be delivered.

All amateurs who handle traffic are invited to report the number of messages they handle to their ARRL Section Manager (listed on page 8 of *QST* every month). For examples of how to arrive at the totals and for more information on traffic handling, consult *The ARRL Operating Manual*.

The ARRL Field Organization

Most of the work of the League is done by volunteers—people just like you who are willing to accept some responsibility for what happens in Amateur Radio.

The League isn't only in Newington, Connecticut. It's everywhere. It's *you*.

WHY THE LEAGUE WAS FORMED

The story of the founding of the American Radio Relay League in 1914 is told in Clinton B. DeSoto's (W1CBD) classic history of Amateur Radio's first three decades, *Two Hundred Meters and Down*. In a nutshell, the League grew from Hiram Percy Maxim's (1AW) vision of a highly qualified network of amateur stations organized to relay messages from one station to another. In the beginning, even the paying of dues was voluntary. More important to maintaining one's status as a "Relay Station" was consistent activity and the observing of high operating standards, a tradition which has been preserved to this day in the programs of the League's Field Services Department.

This is our heritage, as radio amateurs and as members of the League. We owe a great deal to our predecessors who were willing to sacrifice their time and money for a cause they believed in: The preservation and promotion of Amateur Radio. What about today, and the future? How do we protect and expand the privileges we now enjoy? What will be our legacy to future generations of radio amateurs? That depends on our willingness to make the same sacrifices and to support a unified organization that can speak with authority on our behalf. Some of the work can be done most effectively by a paid, professional staff; this is where Newington comes into the picture and why your ARRL membership dues are important. But the biggest job we face is at the local level: Public relations, training new amateurs, organizing emergency communications capabilities, heading off unreasonable restrictions on antennas, solving TVI and other interference problems, building healthy relationships with local officials and opinion leaders, and so on. The responsibility for solving local problems must rest at the local level: As amateurs we could never afford to hire a professional staff large enough to handle them all on our behalf. The importance of member participation in promoting and preserving Amateur Radio has never been greater.

ENTER THE ARRL FIELD ORGANIZATION

The Headquarters could never begin to accomplish the amount of critical work performed by dedicated volunteers in the field. From emergency communications, to public relations, to government liaison, to message handling, to club roles in the community, to rules compliance monitoring, to technical advancement, to information dissemination and more, the Field Organization comes through for Amateur Radio. To find your niche, take a look at the following job descriptions; there's one that will fit you perfectly.

The ARRL Field Organization is divided for administrative purposes into geographical units called sections. There are sixty-nine sections throughout the US; some states have more than one section. The chief-of-staff is the *Section Manager (SM)* who heads a team of hand-picked assistants who look after their particular specialties. These *Section level Appointees* also look after their flock of *Station Appointees* who carry out duties at the grass-roots level. Section Managers are elected by the section ARRL membership, while others in the Field Organization are appointed by the SM or his designee(s).

Station Appointments

You may apply for a station appointment; please use the application form at the back of this booklet.

Official Relay Stations handle traffic and develop communications skills necessary in emergencies.

Official Emergency Stations set the example for others in providing emergency communications capabilities. These stations can operate without commercial power if necessary.

Official Observers monitor the bands and advise amateurs of operating discrepancies. In hard-core rules violation cases, they are trained to gather evidence for possible FCC enforcement action.

Official Bulletin Stations manage information throughout the section. They receive and disseminate ARRL and other bulletins via packet bulletin boards, phone and CW nets, and other media.

Assistant Technical Coordinators serve as local technical resources to clubs and individual hams. They help solve RFI problems and answer technical questions.

Public Information Officers provide local public information and public relations, especially at the club level. They develop media relations through which the community is informed of Amateur Radio activities.

Local Government Liaisons are primarily responsible for monitoring proposals and actions by local government bodies and officials which may affect Amateur Radio. They work with the local PIO to alert section leadership officials and area amateurs encountering problems dealing with local government agencies, and coordinate local responses.

Section Appointments

Above the station level are appointees who help the Section Manager manage the section in their specific areas of expertise. *Section Traffic Managers*, *Section Emergency Coordinators*, *Affiliated Club Coordinators*, *Bulletin Managers*, *Net Managers*, *Public Information Coordinators*, *State Government Liaisons*, *Technical Coordinators*, and *Official Observer Coordinators* are experts in their field. They coordinate the efforts of their station appointees under authority delegated to them by the SM.

National Traffic System

Relaying and delivering noncommercial "greeting" messages for the public is what the National Traffic System (NTS) is all about. It's a great public service and NTS has the excellent fringe benefit of training amateur operators in proper message handling techniques for use in possible emergency communications situations.

The National Traffic System consists of four different levels of nets that meet in an orderly time sequence to provide a flow pattern for traffic from point of origin to point of destination. A message flows through the National Traffic System in a manner similar to an airline passenger who starts out in a small residential town with a destination across the continent in another small town. He has to change carriers many times in the process, starting with a local ground conveyance to a feeder air line, to a transcontinental air line, to another feeder air line, then local transportation to deliver him to his destination. In a similar manner, the transcontinental message starts with the originating station in a local net, is carried to the section net, the region net, the area net, via Transcontinental Corps (TCC) to a distant area net, then back down the line to delivery.

Of course the message, like the passenger, can "get on" or "get off" at any point if that's its origin or destination. Thus, a message from, say, New York to Detroit would never get on TCC, but would "get off" at area level. A message from San Francisco to Los Angeles would not go beyond region level, and one from San Antonio to Houston would remain inside the section net.

Local nets are those that cover small areas such as a community, city, county or metropolitan area, but not a complete ARRL section. They usually operate on VHF (typically 2-meter FM) at times and on days most convenient to their members. Some local nets are designated as "emergency" (ARES) nets that do not specialize in traffic handling. Local nets are intended mainly for local delivery of traffic on an emergency basis: There's not much to be gained by sending ordinary messages over radio when they could be delivered conveniently by a local telephone call. Some NTS local nets operate on a daily basis, just as do other nets of the system, to put the traffic as close as possible to its actual destination before delivery—a matter of practice in a procedure that might be required in an emergency.

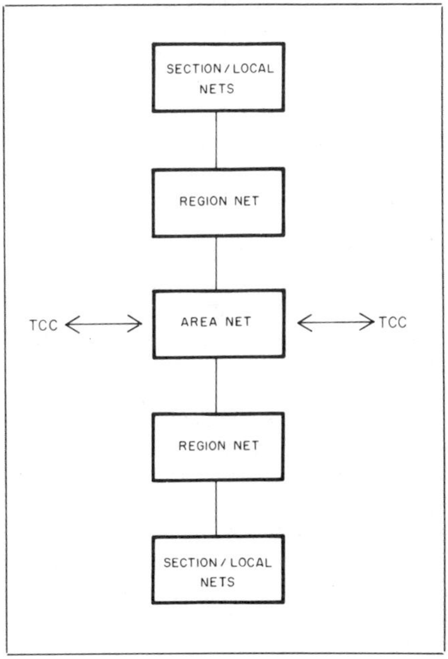

NTS SEQUENCING

PACKET-BASED TRAFFIC HANDLING

In recent times, a new system of traffic handling for the public has emerged, providing a powerful supplement to the more traditional CW- and phone-based operations of NTS. Based on packet radio, the popular bulletin board systems, and the growing transcontinental network of packet message *autoforwarding nodes*, the system is handling more and more traffic daily. It offers advanced reliability, error-free transmission, and speed.

It works much along the same lines as the traditional NTS. A message is uploaded to a local section traffic node (a neighborhood packet bulletin board supporting NTS operation) by a traffic handler. The node then checks its autoforwarding table for information on where to route the message, then sends it on its way towards its ultimate destination. A local traffic handler at the other end then downloads the message and either delivers it personally, or brings it to a local or section phone or CW net for relay to a station that can make the delivery.

Packet Bulletin Board System NTS Traffic Commands

Handling traffic via packet radio? It's simple. Follow these basic steps:

• Log on to your local NTS Section Node (PBBS).

• *Listing Radiogram Traffic.* Type the command LT (*List Traffic*). The BBS will sort and display an index of all radiograms available for handling, something like this:

MSG#	TR	SIZE	TO	FROM	@BBS	DATE	Title
200	TN	282	06105	KY1T	NTSCT	870225	QTC 1 Hartford
198	TN	302	02973	K1CE	NTSRI	870224	QTC 1 Cranston
192	TN	215	46215	WF4B	NTSIN	870224	QTC 1 Indianapolis
190	TN	200	06716	AJ6F	NTSCT	870224	QTC 1 Waterbury
188	TN	315	06111	KH6WZ		870224	QTC 1 Newington
187	TN	320	NTSCT	K6TP		870224	QTC 1 New Haven
186	TN	300	NTSAL	WA4STO		870224	QTC 1 Birmingham
184	TN	295	NTSCA	WB4FDT		870224	QTC 1 Fresno

• *Receiving (downloading) Radiograms.* To take a message off the bulletin board for telephone delivery to the third party, or for relay to an NTS local or section net, type the R or RT (*Read Traffic*) command and the message number. For example, R 188 will cause the BBS to find and send the message text file containing the radiogram for the third party in Newington. The radiogram will look like any other, with preamble, address, text and signature. There is a slight difference: Some additional packet-related message header information is added. This information includes the routing path of the message so readers can see if any problems caused delays in the system.

• After the message is saved to your printer or disk, the message should be removed from the PBBS ("killed") by using the KT (*Kill Traffic*) command and the message number. For example, type KT 188 at the BBS prompt to kill message 188. This procedure saves the message from being delivered twice.

• *Delivering or Relaying Messages.* A downloaded radiogram should, of course, be handled expeditiously in the traditional way: telephone delivery, or relay to a phone or CW net.

• *Sending (uploading) Messages.* To send a radiogram, use the ST (*Send Traffic*) command. The BBS will prompt you for where the message should be routed (type in the

ZIP code, or section designator), the message title which contains the city in the address of the radiogram (Dayton), and the text of the message in radiogram format. The BBS, usually within the hour, will check its outgoing mailpouch, find the NTSOH or ZIP-addressed message and automatically forward it to the next packet station in line to the NTSOH node.

WE WANT YOU

Local and section BBSs need to be checked daily for NTS traffic. System operators and STMs can't do it alone. They need your help to clear NTS radiograms every day, seven days a week, for delivery and relay. If you are a traffic handler/packeteer, contact your Section Traffic Manager or Section Manager for information on existing NTS/packet procedures in your section.

If you are a packeteer and know nothing of NTS traffic handling, contact ARRL HQ, your Section Manager or Section Traffic Manager for information on how you can put your packet radio gear to use in serving the public in routine times, but especially in times of emergency!

Amateur Radio Emergency Service

The Amateur Radio Emergency Service (ARES) consists of licensed amateurs who have voluntarily registered their qualifications and equipment for communication duty in the public service when disaster strikes. Every licensed amateur, whether or not a member of ARRL or any other local or national organization, is eligible for membership in the ARES. The only qualification other than holding a license is a sincere desire to serve. Since ARES is an amateur-provided service, only amateurs are eligible for membership. The possession of emergency-powered equipment is desirable, but is not a requirement for membership.

PRINCIPLES OF EMERGENCY COMMUNICATION

It is impossible to state exact rules to try to cover every situation that will arise. The good average amateur faced with an emergency situation may, however, benefit greatly from certain rules of thumb. These rules are, or should be, part of his training in his ARES group. They are presented here somewhat at random and should be digested by all amateurs, whether active in emergency communications preparation or not.

1) *Keep the QRM level down.* In an emergency, many of the most crucial stations will be weak in signal strength. It is most essential that all other stations remain silent unless they are called upon. When in doubt as to whether or not to turn on your transmitter, *don't*. Our amateur bands are very congested. If you want to help, study the situation by listening. Don't transmit unless you are sure you can help by doing so. Don't ever break into an emergency net just to inform the control station you are there if needed.

2) *Monitor established emergency frequencies.* Many localities and some geographical areas have established emergency frequencies on which someone is always (or nearly always) monitoring for possible calls. When not otherwise engaged, it is helpful simply to sit and listen on such frequencies, some of which are used for general rag chewing as well as emergency preparedness drilling. On CW, SOS (without spaces) is universally recognized but has some legal aspects that should be considered where the need is not truly crucial. On voice, one can use "MAYDAY" (universal, the phone equivalent of SOS) or, to break into a net or conversation, the word "emergency."

3) *Avoid spreading rumors.* During an emergency, especially on the phone bands, you may hear almost anything. Unfortunately, much misinformation is transmitted. Rumors are started by expansion, deletion, amplifying or modifying words, exaggeration or interpretation. All addressed transmissions should be officially authenticated as to their source. They should be repeated word for word, if at all, and only when specifically authorized. In an emergency situation, with everyone's nerves on edge, it is little short of criminal to make a statement on the air without foundation in authenticated fact.

4) *Authenticate all messages.* Every message which purports to be of an official nature should be written

and signed. Whenever possible, amateurs should avoid initiating emergency traffic themselves. We do the communicating; served-agency officials supply the content of the communications.

5) *Strive for efficiency.* Whatever happens in an emergency, you will find hysteria, and some amateurs who are activated by the thought that they must be "sleepless heroes." Instead of operating your own station full time at the expense of your health and efficiency, it is much better to serve a shift at one of the best located and best equipped stations, suitable for the work at hand, manned by relief shifts of the best-qualified operators. This reduces interference and secures well-operated stations.

6) *Select the mode and band to suit the need.* It is a characteristic of all amateurs to believe that their favorite mode and band is superior to all others. For certain specific purposes and distances, one or the other may indeed be preferable. The merits of the use of each in a communications emergency should be evaluated impartially with a view to apportioning to each the type of work for which it is best suited. There is, of course, no alternative to using what happens to be available, but there are right and wrong purposes for which it might be used. Long experience has developed the following advantages:

For CW:

1) Less QRM in most amateur bands.

2) Secrecy of communications—contents of communications cannot be intercepted, generally speaking, by the general public, to start rumors or undue concern.

3) Simpler transmitting equipment.

4) Greater accuracy in record communications.

5) Longer range for a given

KB3OM came across this devastation as part of a team surveying damage after tornadoes moved through Lycoming County, Pennsylvania.

amount of power.

For Phone:

1) More practical for portable and mobile work.

2) More widespread availability of operators.

3) Faster communication for tactical or "command" purposes.

4) More readily appreciated and understood by the public.

5) Official to official and phone patch communication.

For Packet:

Advantages (1) and (2) of CW, advantage (2) of phone, plus greater speed in record communication than any other mode.

The well-balanced emergency organization will have CW, phone and packet modes available, to utilize the advantages of all three as outlined above. Of course, one must make the best use of whatever is available, but a great deal of efficiency is lost when there is lack of coordination or liaison between the different types of operation in an emergency. Absolute impartiality, willingness to let performance speak for itself—these are prime requisites if we are to realize the best possible results.

7) *Use all communications channels intelligently.* The object of emergency communications is to save lives—anything else is incidental. Amateur Radio is a secondary communications means; normal channels are primary and should be used if available. Emergency channels other than Amateur Radio that are available in the absence of amateur channels should be utilized without fear of favoritism in the interest of getting the message through.

8) *Don't "broadcast."* Some amateur stations in an emergency situation have a tendency to emulate "broadcast" techniques. While it is true that the general public may listen to many of our transmissions, they are not and should not be made for that purpose. Broadcast stations are well-equipped to perform any such service. Our job is to communicate for, not with, the general public. FCC regulation 97.113 says amateur stations may not be used to disseminate radio communications intended to be received by the general public and may not be used for any activity directly related to program production or newsgathering for broadcast purposes. The retransmission of programs or signals emanating from any class of radio station other than amateur, except for emergency communications, is prohibited.

APPENDIX

JOIN ARRL TODAY!

A bona fide interest in Amateur Radio is the only essential requirement, but full voting membership is granted only to licensed radio amateurs of the US. Therefore, if you have a license, please be sure to indicate it below. Please print.

☐ New Member ☐ Previous Member ☐ Renewal

____

Class of License Call Sign Date of Application

____

Name

____

Address

____

City, State, ZIP

Membership Class	☐ Regular		☐ Family ☐ Blind	☐ 65 or older	
	US	ELSEWHERE*		US	ELSEWHERE*
☐ 1 year	$30	$ 42	$ 4	$24	$ 36
☐ 2 years	57	81	8	45	69
☐ 3 years	80	116	12	65	101

*These rates include the postage surcharge which partially offsets the additional cost to mail *QST* outside the US.

IMPORTANT: Please attach your Expiration Notice to this form if you are renewing your current membership. Payment in US Funds only.

A member of the immediate family of a League member, living at the same address, may become a League member without *QST* at the special rate of $4 per year. Family membership must run concurrently with that of the member receiving *QST*. Blind amateurs may join without *QST* for $4 per year.

Persons who are age 65 or older may, upon request, apply for League membership at the reduced rates shown. A one-time proof of age, in the form of a driver's license or birth certificate, is required. If you are 17 or younger, a special rate may apply.

Your membership certificate will be mailed to you about 2 weeks from the date we receive your application. Delivery of *QST* may take slightly longer, but future issues should reach you on a regular basis. Membership is available only to individuals. Fifty percent of dues is allocated to *QST* and the balance for membership.

DUES ARE SUBJECT TO CHANGE WITHOUT NOTICE.

THE AMERICAN RADIO RELAY LEAGUE, INC

225 MAIN STREET NEWINGTON, CONNECTICUT 06111 USA

ARRL Field Organization

The United States is divided into 15 ARRL Divisions. Every two years, the ARRL full members in each of these divisions elect a director and a vice director to represent them on the League's Board of Directors. The Board determines the policies of the League, which are carried out by the Headquarters staff. A director's function is principally policymaking at the highest level, but the Board of Directors is all-powerful in the conduct of League affairs.

The 15 divisions are further broken down into 69 sections and the ARRL full members in each section elect a Section Manager (SM). The SM is the senior elected ARRL official in the section, and in cooperation with the director, fosters and encourages all ARRL activities within the section. A breakdown of sections within each division (and counties within each split-state section) follows:

ATLANTIC DIVISION: *Delaware; Eastern Pennsylvania* (Adams, Berks, Bradford, Bucks, Carbon, Chester, Columbia, Cumberland, Dauphin, Delaware, Juniata, Lackawanna, Lancaster, Lebanon, Lehigh, Luzerne, Lycoming, Monroe, Montgomery, Montour, Northampton, Northumberland, Perry, Philadelphia, Pike, Schuylkill, Snyder, Sullivan, Susquehanna, Tioga, Union, Wayne, Wyoming, York); *Maryland-D.C.; Southern New Jersey* (Atlantic, Burlington, Camden, Cape May, Cumberland, Gloucester, Mercer, Ocean, Salem); *Western New York* (Allegany, Broome, Cattaraugus, Cayuga, Chautauqua, Chemung, Chenango, Clinton, Cortland, Delaware, Erie, Essex, Franklin, Fulton, Genesee, Hamilton, Herkimer, Jefferson, Lewis, Livingston, Madison, Monroe, Montgomery, Niagara, Oneida, Onondaga, Ontario, Orleans, Oswego, Otsego, St Lawrence, Schoharie, Schuyler, Seneca, Steuben, Tioga, Tompkins, Wayne, Wyoming, Yates); *Western Pennsylvania* (those counties not listed under Eastern Pennsylvania).

CENTRAL DIVISION: *Illinois; Indiana; Wisconsin.*

DAKOTA DIVISION: *Minnesota; North Dakota; South Dakota.*

DELTA DIVISION: *Arkansas; Louisiana; Mississippi; Tennessee.*

GREAT LAKES DIVISION: *Kentucky; Michigan; Ohio.*

HUDSON DIVISION: *Eastern New York* (Albany, Columbia, Dutchess, Greene, Orange, Putnam, Rensselaer, Rockland, Saratoga, Schenectady, Sullivan, Ulster, Warren, Washington, Westchester); *NYC-LI* (Bronx, Kings, Nassau, New York, Queens, Staten Island, Suffolk); *Northern New Jersey* (Bergen, Essex, Hudson, Hunterdon, Middlesex, Monmouth, Morris, Passaic, Somerset, Sussex, Union, Warren).

MIDWEST DIVISION: *Iowa; Kansas; Missouri; Nebraska.*

NEW ENGLAND DIVISION: *Connecticut; Maine, Eastern Massachusetts* (Barnstable, Bristol, Dukes, Essex, Middlesex, Nantucket, Norfolk, Plymouth, Suffolk); *New Hampshire; Rhode Island; Vermont; Western Massachusetts* (those counties not listed under Eastern Massachusetts).

NORTHWESTERN DIVISION: *Alaska; Idaho; Montana; Oregon; Eastern Washington* (Okanogan, Chelan, Douglas, Kittitas, Yakima, Klickitat, Benton, Franklin, Walla Walla, Adams, Grant, Lincoln, Ferry, Stevens, Pend Oreille, Spokane, Whitman, Columbia, Garfield and Asotin; *Western Washington* (Whatcom, Skagit, San Juan, Snohomish, King, Pierce, Thurston, Clallam, Jefferson, Mason, Kitsap, Grays Harbor Island, Pacific, Lewis, Wahkiakum, Cowlitz, Clark and Skamania).

PACIFIC DIVISION: *East Bay* (Alameda, Contra Costa, Napa, Solano); *Nevada; Pacific* (Hawaii and US possessions in the Pacific); *Sacramento Valley* (Alpine, Amador, Butte, Colusa, El Dorado, Glenn, Lassen, Modoc, Nevada, Placer, Plumas, Sacramento, Shasta, Sierra, Siskiyou, Sutter, Tehama, Trinity, Yolo, Yuba); *San Francisco* (Del Norte, Humboldt, Lake, Marin, Mendocino, San Francisco, Sonoma); *San Joaquin Valley* (Calaveras, Fresno, Kern, Kings, Madera, Mariposa, Merced, Mono, San Joaquin, Stanislaus, Tulare, Tuolumne); *Santa Clara Valley* (Monterey, San Benito, San Mateo, Santa Clara, Santa Cruz).

ROANOKE DIVISION: *North Carolina; South Carolina; Virginia; West Virginia.*

ROCKY MOUNTAIN DIVISION: *Colorado; Utah; New Mexico; Wyoming.*

SOUTHEASTERN DIVISION: *Alabama; Georgia; Northern Florida* (Alachua, Baker, Bay, Bradford, Calhoun, Citrus, Clay, Columbia, Dixie, Duval, Escambia, Flagler, Franklin, Gadsden, Gilchrist, Gulf, Hamilton, Hernando, Holmes, Jackson, Jefferson, Lafayette, Lake, Leon, Levy, Liberty, Madison, Marion, Nassau, Okaloosa, Orange, Pasco, Putnam, Santa Rosa, Seminole, St Johns, Sumter, Suwanee, Taylor, Union, Volusia, Wakulla, Walton, Washington); *Southern Florida* (those counties not listed under Northern Florida); *Puerto Rico; US Virgin Islands.*

SOUTHWESTERN DIVISION: *Arizona; Los Angeles; Orange* (Inyo, Orange, Riverside, San Bernardino); *San Diego* (Imperial, San Diego); *Santa Barbara* (San Luis Obispo, Santa Barbara, Ventura).

WEST GULF DIVISION: *North Texas* (Anderson, Archer, Baylor, Bell, Bosque, Bowie, Brown, Camp, Cass, Cherokee, Clay, Collin, Comanche, Cooke, Coryell, Dallas, Delta, Denton, Eastland, Ellis, Erath, Falls, Fannin, Franklin, Freestone, Grayson, Gregg, Hamilton, Harrison, Henderson, Hill, Hood, Hopkins, Hunt, Jack, Johnson, Kaufman, Lamar, Lampasas, Limestone, McLennan, Marion, Mills, Montague, Morris, Nacogdoches, Navarro, Palo Pinto, Panola, Parker, Rains,

Red River, Rockwall, Rusk, Shelby, Smith, Somervell, Stephens, Tarrant, Throckmorton, Titus, Upshur, Van Zandt, Wichita, Wilbarger, Wise, Wood, Young); *Oklahoma; South Texas* (Angelina, Aransas, Atacosa, Austin, Bandera, Bastrop, Bee, Bexar, Blanco, Brazoria, Brazos, Brooks, Burleson, Burnet, Caldwell, Calhoun, Cameron, Chambers, Colorado, Comal, Concho, DeWitt, Dimmitt, Duval, Edwards, Fayette, Fort Bend, Frio, Galveston, Gillespie, Goliad, Gonzales, Grimes, Guadalupe, Hardin, Harris, Hays, Hidalgo, Houston, Jackson, Jasper, Jefferson, Jim Hogg, Jim Wells, Karnes, Kendall, Kenedy, Kerr, Kimble, Kinney, Kleberg, LaSalle, Lavaca, Lee, Leon, Liberty, Live Oak, Llano, Madison, Mason, Matagorda, Maverick, McCulloch, McMullen, Medina, Menard, Milam, Montgomery, Newton, Nueces, Orange, Polk, Real, Refugio, Robertson, Sabine, San Augustine, San Jacinto, San Patricio, San Saba, Starr, Travis, Trinity, Tyler, Uvalde, Val Verde, Victoria, Walker, Waller, Washington, Webb, Wharton, Willacy, Williamson, Wilson, Zapata, Zavala); *West Texas* (Andrews, Armstrong, Bailey, Bordon, Brewster, Briscoe, Callahan, Carson, Castro, Childress, Cochran, Coke, Coleman, Collingsworth, Cottle, Crane, Crockett, Crosby, Culberson, Dallam, Dawson, Deaf Smith, Dickens, Donley, Ector, El Paso, Fischer, Floyd, Foard, Gaines, Garza, Glasscock, Gray, Hale, Hall, Hansford, Hardeman, Hartley, Haskell, Hemphill, Hockley, Howard, Hudspeth, Hutchinson, Irion, Jeff Davis, Jones, Kent, King, Knox, Lamb, Lipscomb, Loving, Lubbock, Lynn, Martin, Midland, Mitchell, Moore, Motley, Nolan, Ochiltree, Oldham, Parmer, Pecos, Potter, Presidio, Randall, Reagan, Reeves, Roberts, Runnels, Schleicher, Scurry, Shackelford, Sherman, Sterling, Stonewall, Sutton, Swisher, Taylor, Terrell, Terry, Tom Green, Upton, Ward, Wheeler, Winkler, Yoakum).

CRRL-CANADA: *Alberta; British Columbia; Manitoba; Maritime* (Nova Scotia, New Brunswick, Prince Edward Island, Labrador, Newfoundland); *Ontario; Quebec; Saskatchewan.*

ARRL FIELD ORGANIZATION
APPLICATION FOR STATION APPOINTMENT

I, ______________________________, licensee of Amateur Radio station ____________, hereby submit application to the Section Manager (or designated Section-level official) of the ____________________ Section, for appointment as:

Official Relay Station ______
Official Emergency Station ______
Official Observer ______ (Must hold minimum Technician Class license for 4 or more years).
Official Bulletin Station ______
Assistant Technical Coordinator ______
Public Information Officer ______
Local Government Liaison ______

My qualifications are:

____
(describe appropriate experience and/or interest)
____
____
____

Address: ______________________________
street city state zip

Phone number: ______________________________
work home

ARRL membership number: ______________________________ (from *QST* label)

I have read the description of the duties and responsibilities of the appointment, and agree to maintain current League membership, and report my station activity to designated section officials on a monthly basis.

Signed: ______________________________ Date: ____________

(FSD-187 10/90)

ARRL PUBLICATIONS

THE 1991 ARRL HANDBOOK

This is the most comprehensive edition since the *Handbook* was first published in 1926. It is updated yearly to present the cutting edge of RF communication techniques while presenting hundreds of projects the average Amateur Radio operator can build. This publication belongs in every ham shack.

Hardcover only . . . . **#1689 $25**

ANTENNA BOOKS

THE ARRL ANTENNA BOOK represents the best and most highly regarded information on antenna fundamentals, transmission lines, design and construction of wire antennas as well as Yagis and quads for HF. The 15th edition has over 700 pages of practical antenna information.

©1988, Softcover . . . . **#2065 $18**

Novice Antenna Notebook is written for the beginner or experienced amateur who wants practical information on basic antenna designs and construction.

©1988. Softcover . . . . **#3088 $10**

W1FB's Antenna Notebook Practical wire and vertical antenna designs . . **#2618 $10**

Antenna Compendium
©1985 178 pages . . . . **#0194 $10**

Antenna Compendium Vol II . **#2545 $12**
Packed with new material on quads, Yagis and other interesting topics.

PASSING POWER!—THESE PUBLICATIONS WILL HELP YOU THROUGH THE EXAMS

Beginning with **Tune in the World with Ham Radio** for the Novice and progressing through the critically acclaimed **ARRL License Manual Series** for the Technician through Extra Class, you will find passing each exam element a snap! The latest edition of **The FCC Rule Book** is invaluable as a study guide for the regulatory material found on the exams and as a handy reference. *Every* amateur needs an up-to-date copy.

Tune in the World with Ham Radio
Kit with book and cassettes . . . **#2472 $19**
Book only . . . . **#2464 $14**

First Steps in Radio from *QST* presents electronic principles for the beginner.
#2286 $5

License Manual Series
Technician Class . . . . **#2375 $6**
General Class . . . . **#2383 $6**
Advanced Class . . . . **#3274 $6**
Extra Class . . . . **#3282 $8**
FCC Rule Book . . . . **#2456 $9**

ARRL Code Practice Cassettes Each set includes two 90-minute tapes with three hours of Morse code practice consisting of words, sentences, random code groups, related text and practice QSOs.
Set 1—5-10 WPM . . . . **#2227 $10**
Set 2—10-15 WPM . . . . **#2235 $10**
Set 3—15-22 WPM . . . . **#2243 $10**
Set 4—13-14 WPM . . . . **#2251 $10**

OPERATING

The ARRL Operating Manual 688 pages packed with information on how to make the best use of your station, including interfacing home computers, OSCAR, VHF-UHF, contesting, DX, traffic emergency matters and shortwave listening.

©1987 3rd ed. . . . . **#1086 $15**

The ARRL Repeater Directory . **#2987 $6**

The ARRL Net Directory—free shipping
#0275 $1

PACKET RADIO/COMPUTERS

Your Gateway to Packet Radio All you need to know about how to get started, equipment you need and more . . . . **#2030 $12**

DX

The Complete DX'er by W9KNI . **#2083 $12**

The DXCC Companion . **#3398 $6**

DXCC Countries List—free shipping
#0291 $1

Low Band DXing ©1987 . . . . **#047X $10**

CALLBOOKS—1991

North American Callbook . . . **#C091 $30**

International Callbook . . . . . . **#C191 $30**

QRP

QRP Notebook by Doug DeMaw, W1FB. An exciting book for the low-power enthusiast. Includes theory and simple projects.
#3061 $6

SPACE COMMUNICATIONS

The Satellite Experimenter's Handbook by Martin Davidoff, K2UBC . . . . . . **#0046 $10**

Satellite Anthology . . . . **#2103 $8**

ADVENTURE

Grand Canyon QSO (Tompkins)**#5048 $5**
SOS at Midnight (Tompkins)...**#5005 $5**
CQ Ghost Ship (Tompkins)....**#5013 $5**
DX Brings Danger (Tompkins)..**#5021 $5**
Death Valley QTH (Tompkins)..**#503X $5**
Murder by QRM (Tompkins)....**#5064 $5**
Set of 6 Tompkins books....**#1490 $25**
Night Signals (Wall)..........**#2588 $5**

Prices are subject to change without notice. Payment must be in US funds drawn on a bank within the US.

———————— USE THIS FORM OR PHOTOCOPY ————————

ORDER FORM: Please allow 1 week for us to receive your order, 1 week for processing and 1 to 3 weeks shipping time in the US after your order leaves the ARRL. In the US, add the following amounts to your order to cover shipping and handling. Add an additional $1.50 to the mail rate for shipment via surface mail outside the US, call or write for airmail rates. Include street address for UPS. Prices are subject to change without notice.

Amount of Order	Mail	UPS	Amount of Order	Mail	UPS
Less than $20.00	$2.50	$3.50	$40.01—$50.00	$5.50	$6.50
$20.01—$30.00	3.50	4.50	50.01—$75.00	6.50	7.50
30.01— 40.00	4.50	5.50	Over $75.00	7.50	8.50

Product #	Quantity	Title	
		Subtotal for books and nonexempt supplies	
		Enter shipping and handling based on above subtotal	
		Enter items exempt from shipping charges below	
		Donation to the Legal Defense Fund ($1 min.)	
Payment must be made in US Funds drawn on a US bank		TOTAL	

$7.50 minimum on all credit card orders

Charge to ☐VISA ☐MasterCard ☐AMEX ☐Discover

Name ____________________

Call ____________________

Street ____________________

City ____________________

____

____
Card Number

Card good from ____________________

Card good to ____________________
Expiration Date

Signature ____________________

ARRL 225 MAIN STREET NEWINGTON, CT 06111 USA

ABBREVIATIONS, PROSIGNS, PROWORDS

CW	PHONE (meaning or purpose)
AA	(Separation between parts of address or signature.)
AA	All after (use to get fills).
AB	All before (used to get fills).
ADEE	Addressee (name of person to whom message addressed).
ADR	Address (second part of message).
AR	End of message (end of record copy).
ARL	(Used with "check," indicates use of ARRL numbered message in text.)
AS	Stand by; wait.
B	More (another message to follow).
BK	Break; break me; break-in (interrupt transmission on cw. Quick check on phone).
BT	Separation (break) between address and text; between text and signature.
C	Correct; yes.
CFM	Confirm. (Check me on this).
CK	Check.
DE	From; this is (preceding identification).
HH	(Error in sending. Transmission continues with last word correctly sent.)
HX	(Handling instructions. Optional part of preamble.) Initial(s). Single letter (s) to follow.
IMI	Repeat; I say again. (Difficult or unusual words or groups.)
K	Go ahead; over, reply expected. (Invitation to transmit.)
N	Negative, incorrect; no more. (No more messages to follow.)
NR	Number. (Message follow.)
PBL	Preamble (first part of message).
......	Read back. (Repeat as received.)
R	Roger; point. (Received; decimal point.)
SIG	Signed; signature (last part of message).
SK	Out; clear (end of communication, no reply expected).
TU	Thank you.
WA	Word after (used to get fills).
WB	Word before (Used to get fills).
......	Speak slower.
......	Speak faster.

INTERNATIONAL Q SIGNALS

A Q signal followed by a ? asks a question. A Q signal without the ? answers affirmatively, unless otherwise indicated.

QRA	What is the name of your station?
QRG	What's my exact frequency?
QRH	Does my frequency vary?
QRI	How is the tone of my transmission? (1-3)
QRK	What is my signal intelligibility? (1-5)
QRL	Are you busy?
QRM	Is my transmission being interfered with?
QRN	Are you troubled by static?
QRO	Shall I increase transmitter power?
QRP	Shall I decrease transmitter power?
QRQ	Shall I send faster?
QRS	Shall I send slower?
QRT	Shall I stop sending?
QRU	Have you anything for me? (Answer in negative).
QRV	Are you ready?
QRW	Shall I tell..........you're calling him?
QRX	When will you call again?
QRZ	Who is calling me?
QSA	What is my signal strength? (1-5)
QSB	Are my signals fading?
QSD	Is my keying defective?
QSG	Shall I send......messages at a time?
QSK	Can you work breakin?
QSL	Can you acknowledge receipt?
QSM	Shall I repeat the last message sent?
QSO	Can you communicate with......direct?
QSP	Will you relay to......?
QSV	Shall I send a series of V's?
QSW	Will you transmit on......?
QSX	Will you listen for......on......?
QSY	Shall I change frequency?
QSZ	Shall I send each word/group more than once? (Answer, send twice or.......)
QTA	Shall I cancel number......?
QTB	Do you agree with my word count? (Answer negative)
QTC	How many messages have you to send?
QTH	What is your location?
QTR	What is your time?
QTV	Shall I stand guard for you......?
QTX	Will you keep your station open for further communication with me?
QUA	Have you news of......?

ARRL QN SIGNALS FOR CW NET USE

QNA*	Answer in prearranged order.
QNB*	Act as relay Between........and........
QNC	All net stations Copy. I have a message for all net stations.
QND	Net is Directed (controlled by net control station).
QNE*	Entire net stand by.
QNF	Net is Free (not controlled).
QNG	Take over as net control station.
QNH	Your net frequency is High.
QNI	Net stations report in.* I am reporting into the net. (Follow with a list or traffic or QRU).
QNJ	Can you copy me? Can you copy........?
QNK*	Transmit message for.......to.......
QNL	Your net frequency is Low.
QNM*	You are QRMing the net. Stand by.
QNN	Net control station is....... What station has net control?
QNO	Station is leaving the net.
QNP	Unable to copy you. Unable to copy.......
QNQ	Move frequency to......and wait for......to finish handling traffic. Then send him traffic for.......
QNR	Answer........and Receive traffic.
QNS	Following Stations are in the net.* (Follow with list.) Request list of stations in the net.
QNT	I request permission to leave the net for......minutes.
QNU*	The net has traffic for you. Stand by.
QNV*	Establish contact with.......on this frequency. If successful, move to.......and send him traffic for........
QNW	How do I route messages for........?
QNX	You are excused from the net.* Request to be excused from the net.
QNY*	Shift to another frequency (or to.......kHz) to clear traffic with.........
QNZ	Zero beat your signal with mine.

***For use only by the Net Control Station.**

Notes on Use of QN Signals

The QN signals listed above are special ARRL signals for use in amateur cw nets *only*. They are not for use in casual amateur conversation. Other meanings that may be used on other services do not apply. Do not use QN signals on phone nets. *Say it with words*. QN signals need not be followed by a question mark, even though the meaning may be interrogatory.

From ______________________

____

____

Please affix postage. Post Office will not deliver without sufficient postage.

Editor, OARS
American Radio Relay League
225 Main Street
Newington, CT USA 06111

..............please fold and tape..............

FEED BACK

Please use this form to give us your comments about the book, and what improvements you would like to see us make in future editions.

Name ______________________________ Call sign ________________

Daytime Phone (   ) ______________________________ Age ________

Address ____

City, State, ZIP ____

How long have you been licensed? _________________________________